Asian Physics
Olympiad
(1st – 8th)
Problems and Solutions

Asian Physics
Olympiad
(1st – 8th)

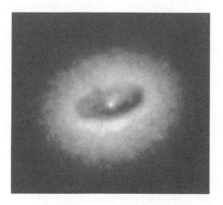

Problems and Solutions

Editor
Zheng Yongling
Fudan University, China

East China Normal University Press

 **World Scientific**

Published by

East China Normal University Press
3663 North Zhongshan Road
Shanghai 200062
China .

and

World Scientific Publishing Co. Pte. Ltd.
5 Toh Tuck Link, Singapore 596224
USA office: 27 Warren Street, Suite 401-402, Hackensack, NJ 07601
UK office: 57 Shelton Street, Covent Garden, London WC2H 9HE

British Library Cataloguing-in-Publication Data
A catalogue record for this book is available from the British Library.

ASIAN PHYSICS OLYMPIAD (1ST–8TH)
Problems and Solutions

ISBN-13 978-981-4271-43-1 (pbk)
ISBN-10 981-4271-43-8 (pbk)

Printed in Singapore by World Scientific Printers

Editor

ZHENG Yongling *Fudan University, China*

Original Authors

National Coaches of the host country of
Asian Physics Olympiad during 2000 – 2007

Copy Editors

NI Ming *East China Normal University Press, China*

ZHANG Ji *World Scientific Publishing Co., Singapore*

ZHAO Junli *East China Normal University Press, China*

Preface

The Asian Physics Olympiad (abbreviated to APhO) is currently the premier physics competition held annually for Asian pre-university or senior high school students. It is modeled after the International Physics Olympiad (IPhO), and demands a similar level of intellectual capability from the participants. The only difference between APhO and IPhO is that each participating country can send eight students at most to compete in APhO instead of five in IPhO. The age of the participants should not exceed twenty on June 30th of the year of the competition.

The idea of creating the Asian Physics Olympiad was first proposed in August 1995 by Dr. Waldemar Gorzkowski, then the President of International Physics Olympiads who regretfully passed away in 2007 during the 38th IPhO held in Isfahhan, Iran. The proposal aimed to promote the quality of science education and attract students to study physics that was much needed in increasing science manpower for developing the new century information economy in Asia region. Technically, APhO was proposed to be held two months before IPhO and it would act as a warm-up competition for the worldwide IPhO. The idea of APhO was welcomed by many Asian countries. Unfortunately, the implement of the proposal was deferred by the Asian financial crisis happened in 1997 through 1998. In 1999, Professor Yohanes Surya with full support from Indonesia government announced to inaugurate the First APhO during the 30th IPhO in Italy. Right after this announcement, Chinese Taipei declared to host the Second APhO in 2001 and was soon followed by Singapore as the host of the Third APhO in 2002 and Thailand as the host of the Fourth APhO in 2003. In the ensuing years, the Fifth to the Ninth APhO were organized smoothly in turn by Vietnam, Indonesia (twice), Kazakhstan, China, and Mongolia from 2004 to 2008, respectively. The number of participating countries has grown from original ten to around twenty. The effect of APhO is very fruitful and conspicuous. The statistical grade data of the past eight years of the global competition of IPhO shows that close to one half of gold medals were won by the students from APhO participating countries.

I am pleased to see the publication of the collection of the APhO problems and solutions. These problems were deliberately formulated by each of the organizing countries. Normally, it had to group together about a dozen of physics professors to form an academic committee and took about one to two years to accomplish the demanding task. Reading and comprehending these problems and solutions can greatly help readers in understanding physics laws deeper and strengthening their analytical and reasoning capability in solving problems. This book is filled with many good wishes and efforts devoted to nourishing our new generations.

Professor Ming-Juey Lin, Ph. D.

Secretary of Asian Physics Olympiads

Contents

Minutes of the First Asian Physics Olympiad

Tangerang-Karawaci (Indonesia), April 24 – May 2, 2000

1. The proposals of the Statutes and Syllabus, prepared by the Organizers and disseminated to all the Asian countries prior to the First Asian Physics Olympiad, have been unanimously accepted. The delegations willing to make changes in the Statutes should send their proposals to the President of the APhO's not later than by December 31, 2000.

2. The following 10 countries were present at the 1st Asian Physics Olympiad: Australia, China, Chinese Taipei, Indonesia, Kazakhstan, Philippines, Singapore, Thailand, Vietnam and Uzbekistan. Australia (non-Asian country) participated as a guest of the Organizers (guest team).

Three countries were represented with observers: Brunei-Darussalam, India and Malaysia.

3. Results of marking the papers by the organizers were presented:

The best score (44.75 points) was achieved by Song Jun-liang from China (Absolute winner of the 1st APhO). The second and third were Kuang Ting Chen (Chinese Taipei)—42.70 points and Zhang Chi (China)—41.75 points.

The following limits for awarding the medals and the honorable mention were established according to the Statutes:

Gold Medal:	38 points,
Silver Medal:	33 points,
Bronze Medal:	27 points,
Honorable Mention:	21 points.

According to the above limits 8 Gold medals, 9 Silver medals, 11 Bronze medals and 17 honorable mentions were awarded. The list of the scores of the winners and the students awarded with honorable mentions were distributed to all the delegations.

4. In addition to the regular prizes a number of special prizes were awarded:

● for the Absolute Winner: Song Jun-liang (China)

● for the best team (prize created by the Director of UNESCO Jakarta Office): the Chinese team: Song Jun-liang, Zhang Chi, Chen Xiao Sheng, Wong Fa, and Dong Shi Ying

● for the best female participant (prize created by the Director of UNESCO

Jakarta Office): Dong Shi Ying (China)

● for the youngest participant (prize created by the Director of UNESCO Jakarta Office): Juan Paolo Asis (Philippines).

5. The International Board has unanimously elected Yohanes Surya, Ph. D., the head of the Organizing Committee of the First Asian Physics Olympiad, to the post of *President of the Asian Physics Olympiads* for five years' term (# 15 of the Statutes). Election of the Secretary of the Asian Physics Olympiads has been postponed to the next Olympiad, which will be held in Taipei in 2001.

6. Dr. Waldemar Gorzkowski, for his merits in establishing the Asian Physics Olympiads, has been unanimously awarded the lifelong title *Honorable President of the Asian Physics Olympiads* (# 15 of the Statutes).

7. President of the APhO's presented a list of the organizers of the competitions in the future. It is:

● 2001 - Taipei (invitations disseminated during the 1st APhO)

● 2002 - Singapore (confirmed orally).

8. The International Board expressed deep thanks to Yohanes Surya, Ph. D. and his collaborators for excellent conducting of the competition. The International Board highlighted all the difficulties occurring when the first event is organized and congratulated the organizers for successfully solving all of them.

9. The Opening Ceremony was honored by the presence of Mr. K. H. Abdurrahman Wahid, President of the Republic of Indonesia; Mrs. Megawati Soekarnoputri, Vice-President of the Republic of Indonesia, honored the Closing Ceremony. Both Guests were welcomed with standing ovation.

10. Action on behalf of the organizers of the next Asian Physics Olympiad Prof. Ming-Juey Lin announced that the 2nd Asian Physics Olympiad will be organized in Taipei on April 22 - May 1, 2001 and cordially invited all the participating countries to attend the competition.

Tangerang, Karawaci   May 2, 2000

Dr. Waldemar Gorzkowski   **Dr. Yohanes Surya**
President of the IPhOs,   Head of the Organizing Committee
Honorable President of the APhOs   of the 1st APhO,
                President of the APhOs

Theoretical Competition

April 25, 2000   Time available: 5 hours

Problem 1
Eclipses of the Jupiter's Satellite

A long time ago before scientists could measure the speed of light accurately, O. Römer, a Danish astronomer studied the times of eclipses of the planet Jupiter. He was able to determined the velocity of light from observed periods of a satellite around the planet Jupiter. Fig. 1 - 1 shows the orbit of the earth E around the sun S and one of the satellites M around the planet Jupiter. (He observed the time spent between two successive emergences of the satellite M from behind Jupiter.)

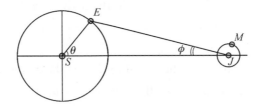

Fig. 1 - 1. The orbits of the earth E around the sun S and a satellite M around Jupiter J. The average distance of the earth E to the Sun is $R_E = 149.6 \times 10^6$ km. The maximum distance is $R_{Emax} = 1.015 R_E$. The period of revolution of the earth is 365 days and of Jupiter is 11.9 years.

A long series of observations of the eclipses permitted an accurate evaluation of the period. The observed period T depends on the relative position of the earth with respect to the frame of reference SJ as one of the axis of coordinates. The average time of revolution is $T_0 = 42$ h 28 m 16 s and maximum observed period is $(T_0 + 15)$ s.

(a) Use Newton's law to estimate the distance of Jupiter to the sun by assuming that the orbits of the earth and Jupiter are circles.

(b) Determine the relative angular velocity ω of the earth with respect to the frame of reference Sun-Jupiter (SJ). Calculate the relative speed of the earth.

(c) Suppose an observer saw M begin to emerge from the shadow when his position was at θ_k and the next emergence when he was at θ_{k+1}, $k = 1$, 2, 3 .... From these observations he got the apparent periods of revolution T (t_k) as a function of time of observations t_k. According to him the variations were due to the variations of the distance of Jupiter $d(t_k)$ relative to the observer during the observations. Derive the distance of Jupiter $d(t_k)$ as a function of time t_k from Fig. 1 - 1 and then use approximate expression to explain how the distance influence observed periods of revolution of M. Estimate the relative error of your approximate distance.

(d) Derive the relation between $d(t_k)$ and $T(t_k)$. Plot period $T(t_k)$ as a function of time of observation t_k. Find the positions of the earth when he observed maximum period, minimum period and true period of the satellite M.

(e) Estimate the speed of light from the above result. Point out sources of errors of your estimation and calculate the order of magnitude of the error.

(f) If the distance of the satellite M to the planet Jupiter $R_M = 4.22 \times 10^5$ km, the distance of the moon M_E to the earth is $R_{ME} = 3.844 \times 10^5$ km and we know that the mass of the earth $= 5.98 \times 10^{24}$ kg and 1 month $= 27$ d 7 h 3 m, find the mass of the planet Jupiter.

Solution

(a) Assume the orbits of the earth and Jupiter are circles, we can write the centripetal force equals gravitational attraction of the Sun.

$$G\frac{M_E M_S}{R_E^2} = \frac{M_E v_E^2}{R_E},$$

$$G\frac{M_J M_S}{R_J^2} = \frac{M_J v_E^2}{R_J},$$

where $G =$ universal gravitational constant,

$M_S =$ mass of the Sun,

$M_E =$ mass of the Earth,

M_J = mass of the Jupiter,

R_E = radius of the orbit of the Earth,

v_E = velocity of the Earth,

v_J = velocity of the Jupiter.

Hence

$$\frac{R_J}{R_E} = \left(\frac{v_E}{v_J}\right)^2.$$

We know

$$T_E = \frac{2\pi}{\omega_J} = \frac{2\pi R_J}{v_J}.$$

We get

$$\frac{T_E}{T_J} = \frac{\dfrac{R_E}{v_E}}{\dfrac{R_J}{v_J}} = \left(\frac{R_E}{R_J}\right)^{\frac{2}{3}},$$

$$R_J = 7.798 \times 10^8 \text{ km}.$$

(b) The relative angular velocity is

$$\omega = \omega_E - \omega_J = 2\pi\left(\frac{1}{365} - \frac{1}{11.9 \times 365}\right)$$

$$= 0.0157 \text{ rad/day},$$

and the relative velocity is

$$v = \omega R_E = 2.36 \times 10^6 \text{ km/day}$$

$$\approx 2.73 \times 10^4 \text{ m/s}.$$

(c) The distance from Jupiter to the Earth can be written as follows:

$$\boldsymbol{d}(t) = \boldsymbol{R}_J - \boldsymbol{R}_E,$$

$$\boldsymbol{d}(t) \cdot \boldsymbol{d}(t) = (\boldsymbol{R}_J - \boldsymbol{R}_E) \cdot (\boldsymbol{R}_J - \boldsymbol{R}_E),$$

$$d(t) = (R_J^2 + R_E^2 - 2R_E R_J \cos \omega t)^{\frac{1}{2}}$$

$$\approx R_J\left[1 - 2\left(\frac{R_E}{R_J}\right)\cos \omega t + \dots\right]^{\frac{1}{2}}$$

$$\approx R_J\left(1 - \frac{R_E}{R_J}\cos \omega t + \dots\right).$$

The relative error of the above expression is of the order

$$\left(\frac{R_E}{R_J}\right)^2 \approx 4\%.$$

The observer saw M begin to emerge from the shadow when his position was at $d(t)$ and he saw the next emergence when his position was at $d(t+T_0)$. Light need time to travel the distance $\Delta d = d(t+T_0) - d(t)$, so the observer will get apparent period T instead of the true period T_0.

$$\Delta d = R_E[\cos \omega t - \cos \omega(t - T_0)]$$
$$\approx R_E \omega T_0 \sin \omega t,$$

since $\omega T_0 \approx 0.03$, $\sin \omega T_0 \approx \omega T_0 + \ldots$, $\cos \omega T_0 \approx 1 - \ldots$.

We can also get this approximation directly from the geometrical relationship from Fig. 1 – 1.

We can also use another method.

From Fig. 1 – 2, we get

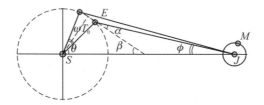

Fig. 1 - 2.   Geometrical relationship to get $\Delta d(t)$.

$$\beta = (\phi + \alpha),$$
$$\frac{\omega_0 T}{2} + \beta + \theta = \frac{\pi}{2},$$

$$\Delta d(t) \approx \omega T_0 R_E \cos \alpha,$$

$$\Delta d(t) \approx \omega T_0 R_E \sin\left(\omega t + \frac{\omega T_0}{2} + \phi\right),$$

$$\omega T_0 \approx 0.03 \text{ and } \phi \approx 0.19.$$

(d) $T - T_0 \approx \dfrac{\Delta d(t)}{c}$; $c =$ velocity of light,

$$T \approx T_0 + \frac{\Delta d(t)}{c} = T_0 + \frac{R_E \omega T_0 \sin \omega t}{c}.$$

The position of the earth when he observed maximum period is at $\omega t = \frac{\pi}{2}$,

and minimum period is at $\omega t = \frac{3}{2}\pi$ and true period at $\omega t = 0$ and π.

(e) From

$$T_{max} = T_0 + \frac{R_E \omega T_0}{c},$$

we get

$$\frac{R_E \omega T_0}{c} = 15.$$

Hence

$$c = 2.78 \times 10^5 \text{ km/s}.$$

We can estimate the relative error that comes from the ratio of spuare of the distance $\left(\frac{R_E}{R_J}\right)^2$ is about 4% and the relative error of the measurement of time is about $\frac{0.5}{15} \times 100\% = 3.4\%$, hence the total relative error is about 7.4%. Another error comes from the assumption that the orbits are circles, actually it is an ellipse. The relative error is about

$$100\% \times \frac{R_{Emax} - R_E}{R_E} \approx 1.5\%.$$

(f) We can calculate the mass of Jupiter by using generalized Kepler's law as in (a), use Newton's law for the moon m circling the earth and the satellite M circling the Jupiter. Hence we get

$$\frac{M_J}{M_E} = \left(\frac{T_{mE}}{T_M}\right)^2 \left(\frac{R_M}{R_{mE}}\right)^3.$$

Hence we get $\quad M_J = 316 M_E = 1.887 \times 10^{27}$ kg.

Problem 2
Detection of Alpha Particles

We are constantly being exposed to radiation, either natural or artificial. With the advance of nuclear power reactors and utilization of radioisotopes in agriculture, industry, biology and medicine, the number of artificial radioactive sources is also increasing every year. One type of radiation emitted by radioactive materials is alpha (α) particle radiation. (An alpha particle is a doubly ionized helium atom having two units of positive charge and four units of nuclear mass.)

The detection of α particles by electrical means is based on their ability to produce ionization when passing through gases and other substances. For an α particle in air at normal (atmospheric) pressure, there is an empirical relation between the mean range R_a and its energy E

$$R_a = 0.318E^{\frac{3}{2}} \tag{1}$$

where R_a is measured in cm and E in MeV.

For monitoring α radiation, one can use an ionization chamber which is a gas-filled detector that operates on the principle of separation of positive and negative charges created by the ionization of gas atoms by the α particle. See Fig. 1 – 3. The collection of charges yields a pulse that can be detected, amplified and then recorded. The voltage difference between anode and cathode is kept sufficiently high so that there is a negligible amount of recombination of charges during their passage to the electrodes.

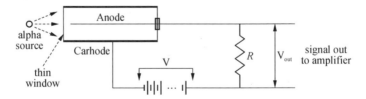

Fig. 1 – 3.   Schematic diagram of ionization chamber circuit.

(a) An ionization chamber electrometer system with a capacitance of 45 picofarad is used to detect α particles having a range R_a of 5.50 cm. Assume the energy required to produce an ion – pair (consisting of a light negative electron and a heavier positive ion, each carrying one electronic charge of magnitude $e = 1.60 \times 10^{-19}$ C) in air is 35 eV. What will be the magnitude of the voltage produced by each α particle?

(b) The voltage pulses such as those due to the α particle of the above problem occur across a resistance R. The smallest detectable saturation current (meaning that the current is more or less constant, indicating that the charge is collected at the same rate at which it is being produced by the incident α particles) with this instrument is 10^{-12} Ampere. Calculate the lowest activity A (disintegration rate of the α emitter radioisotope) of the α source that could be detected by this instrument if the mean range R_a is 5.50 cm assuming a 10% efficiency for the detector.

(c) The above ionization chamber is to be used for pulse counting with a time constant $\tau = 10^{-3}$ seconds. Calculate the resistance and also the necessary voltage pulse amplification required to produce 0.25 V signal.

(d) The ionization chamber has cylindrical plates as shown in Fig. 1 – 4. For a cylindrical counter, the central metal anode and outer thin metal sheath (cathode) have diameters d and D, respectively. Derive the expression for the electric field $E(r)$ and potential $V(r)$ at a radial distance $r\left(\text{with } \dfrac{d}{2} \leqslant r \leqslant \dfrac{D}{2}\right)$ from

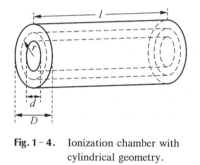

Fig. 1 – 4.   Ionization chamber with cylindrical geometry.

the central axis when the wire carries a charge per unit length λ. Then deduce the capacitance per unit length of the tube. The breakdown field strength of air E_b is 3 MV m^{-1} (breakdown of air occurs at field strengths greater than E_b, maximum electric field in the substance). If $d = 1$ mm and $D = 1$ cm, calculate the potential difference between anode and sheath at which breakdown occurs.

Data: 1 MeV $= 10^6$ eV; 1 picofarad $= 10^{-12}$ F; 1 Ci $= 3.7 \times 10^{10}$ disintegration/second $= 10^6$ μCi (Curie, the fundamental SI unit of activity A); $\displaystyle\int \frac{dr}{r} = \ln r + C$.

 Solution

(a) From the given range – energy relation and the data supplied, we get

$$E = \left(\frac{R_a}{0.318}\right)^{\frac{2}{3}} \text{MeV} = \left(\frac{5.50}{0.318}\right)^{\frac{2}{3}} = 6.69 \text{ MeV}.$$

Since $E_{\text{ion-pair}} = 35$ eV, then

$$N_{\text{ion-pair}} = \frac{6.69 \times 10^6}{35} = 1.9 \times 10^5.$$

We get the size of voltage pulse

$$\Delta V = \frac{\Delta Q}{C} = \frac{N_{\text{ion-pair}} e}{C},$$

with $C = 45$ pF $= 4.5 \times 10^{-11}$ F.

Hence

$$\Delta V = \frac{1.9 \times 10^5 \times 1.6 \times 10^{-19}}{4.5 \times 10^{-11}} \text{ V} = 0.68 \text{ mV}.$$

(b) Electrons from the ion-pairs produced by α particles from a radiaoactive source of mactivity A (= number of α particles emitted by the source per second) which enter the detector with detection efficiency 0.1, will produce a collected current

$$I = \frac{Q}{t} = 0.1 \times AN_{\text{ion-pair}} e$$

$$= 0.1 \times A \times 1.9 \times 10^5 \times 1.6 \times 10^{-19} \text{ A},$$

with $I_{\min} = 10^{-12}$ A,

$$A_{\min} = \frac{10^{-12} \text{dis} \cdot \text{s}^{-1}}{1.6 \times 1.9 \times 10^{-15}} = 330 \text{ dis} \cdot \text{s}^{-1}.$$

Since 1 Ci $= 3.7 \times 10^{10}$ dis $\cdot$ s^{-1}, then

$$A_{\min} = \frac{330}{3.7 \times 10^{10}} \text{ Ci} = 8.92 \times 10^{-9} \text{ Ci}.$$

(c) With time constant

$$\tau = RC \text{ (with } C = 45 \times 10^{-12} \text{ F)} = 10^{-3} \text{ s},$$

$$R = \left(\frac{1000}{45}\right) \text{M}\Omega \approx 22.22 \text{ M}\Omega.$$

For the voltage signal with height $\Delta V = 0.68$ mV generated at the anode of the ionization chamber by 6.69 MeV α particles in problem (a), to achieve a 0.25 V = 250 mV voltage signal, the necessary gain of the voltage pulse amplifier should be

$$G = \frac{250}{0.68} \approx 368.$$

(d) By symmetry, the electric field is directed radially and depends only on distance the axis and can be deducted by using Gauss' theorem. If we construct a Gaussian surface which is a cylinder of radius r and length l, the charge contained within it is σl. See Fig. 1-5.

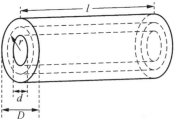

Fig. 1-5. The Gaussian surface used to calculate the electric field E.

The surface integral is

$$\int \boldsymbol{E} \cdot \mathrm{d}\boldsymbol{S} = 2\pi r l E.$$

Since the field E is everywhere constant and normal to the curved surface. By Gauss' theorem:

$$2\pi r l E = \frac{\lambda l}{\varepsilon_0},$$

so

$$E(r) = \frac{\lambda}{2\pi \varepsilon_0 r}.$$

Since E is radial and varies only with r, then $E = -\dfrac{\mathrm{d}V}{\mathrm{d}r}$ and the potential V can be found by integrating $E(r)$ with respect to r. If we call the potential of inner wire V_0, we have

$$V(r) - V_0 = -\frac{\lambda}{2\pi \varepsilon_0} \int_{\frac{d}{2}}^{r} \frac{\mathrm{d}r}{r}.$$

Thus

$$V(r) = V_0 - \frac{\lambda}{2\pi \varepsilon_0} \ln\left(\frac{2r}{d}\right).$$

We can use this expression to evaluate the voltage between the capacitor's conductors by setting $r = \dfrac{D}{2}$, giving a potential difference of

$$V = \frac{\lambda}{2\pi\varepsilon_0} \ln\left(\frac{D}{d}\right).$$

Since the charge Q in the capacitor is σl, and the capacitance C is defined by $Q = CV$, the capcitance per unit length is

$$\frac{2\pi\varepsilon_0}{\ln\dfrac{D}{d}}.$$

The maximum electric field occurs where r minimum, i. e. at $r = \dfrac{d}{2}$. If we set the field at $r = \dfrac{d}{2}$ equal to the breakdown field E_b, our expression for E (r) shows that the charge per unit length σ in the cpacitor must be $E_b \pi \varepsilon_0 d$. Substituting for the potential difference V across the capacitor gives

$$V = \frac{1}{2} E_b d \ln\left(\frac{D}{d}\right).$$

Taking $E_b = 3 \times 10^6$ Vm^{-1}, $d = 1$ mm, and $D = 1$ cm, we have $V = 3.45$ kV.

Problem 3
Stewart-Tolman effect

In 1917, Stewart and Tolman discovered a flow of current through a coil wound around a cylinder rotated axially under angular acceleration.

Consider a great number of rings, each with the radius r, made from a thin metallic wire with resistance R. The rings have been put in a uniform way on a very long glass cylinder which is vacuum inside. Their positions on the cylinder are fixed by gluing the rings to the cylinder. The number of rings per unit of length along the symmetry axis is n. The planes containing the rings are perpendicular to the symmetry axis of the cylinder.

At some moment the cylinder starts a rotational movement around its symmetry axis with a constant angular acceleration α. Find the value of the magnetic field B at the center of the cylinder (after a sufficiently long time). We assume that the electric charge e of an electron, and the electron mass m are known.

Solution

Consider a single ring first.

Let us take into account a small part of the ring and introduce a reference system in which this part is at rest. The ring is moving with certain angular acceleration α. Thus, our reference system is not an inertial one and there exists certain linear acceleration in it. The radial component of this acceleration may be neglected as the ring is very thin and no radial effects should be observed in it. The tangential component of the linear acceleration along the considered part of the ring is $r\alpha$. When we speak about the reference system in which the positive ions forming the crystal lattice of the metal are at rest. In this system certain inertial force acts on the electrons. This inertial force has the value $mr\alpha$ and is oriented in an opposite side to the acceleration mentioned above.

An interaction between the electrons and crystal lattice does not allow electrons to increase their velocity without any limitations. This interaction, according to the Ohm's law, is increasing when the velocity of electrons with respect to the crystal lattice in increasing. At some moment equilibrium between the inertial force and the breaking force due the interaction with the lattice is reached. The next result is that the positive ions and the negative electrons are moving with different velocities; it means that in the system in which the ions are at rest an electric current will flow!

The inertial force is constant and in each point is tangent to the ring. It acts onto the electrons in the same way as certain fictitious electric field tangent to the ring in each point.

Now we shall find value of this fictitious electric field. Of course, the force due to it should be equal to the inertial force. Thus

$$eE = mr\alpha.$$

Therefore

$$E = \frac{mr\alpha}{e}.$$

In the ring (at rest) with resistance R, the field of the above value would generate a current:

$$I = \frac{2\pi rE}{R}.$$

Thus, the current in the considered ring should be:

$$I = \frac{2\pi mr^2\alpha}{R}.$$

It is true that the field E is a fictitious electric field. But it describes a real action of the inertial force onto the electrons. The current flowing in the ring is real!

The above considerations allow us to treat the system described in the text of the problem as a very long solenoid consisting of n loops per unit of length (along the symmetry axis), in which the current I is flowing. It is well known that the magnitude of the field B inside such solenoid (far from its ends) is homogenous and its value is equal:

$$B = \mu_0 nI,$$

where μ_0 denotes the permeability of vacuum. Thus, since the point at the axis is not rotating, it is at rest both in the noninertial and in the laboratory frame, hence the magnetic field at the center at the center of the axis in the laboratory frame is

$$B = \frac{2\pi\mu_0 nmr^2\alpha}{eR}.$$

It seems that this problem is very instructive as in spite of the fact that the rings are electrically neutral, in the system – unexpectedly, due to a specific structure of matter – there occurs a magnetic field. Moreover, it seems that due to this problem it is easier to understand why the electrical term "electromotive force" obtains a mechanical term "force" inside.

Experimental Competition

April 27, 2000   Time available: 5 hours

Problem 1
Determination of the Density of Oil

Listed below are the only apparatus and materials available for your experiment:

(1) Teat tube with uniform cross-section over most of its length between its two ends;

(2) Vessel;

(3) Ruler;

(4) Eye dropper;

(5) Graph papers;

(6) Drying cloth/tissue papers;

(7) Rubber band for level marking;

(8) Distilled water with density 1.00 g/cm^3;

(9) Oil in a plastic cup.

In this experiment, you are to determine the density of the oil without measuring the dimensions of the tube. You should not put both oil and water in the tube at the same time.

Include the following in your report:

(a) The theoretical basis for the analysis;

(b) A description of the method and procedure of the experiment;

(c) Final value for the density of oil;

(d) The errors and their sources.

Solution

Experimental configuration

This experiment is an application of Archimedes' law. The basic experimental configuration is accordingly described by Fig. 1 – 6. It is assumed in this figure that the tube is in an up-right position (perpendicular to water surface). It is also clear that the same reference point must be used for measuring the positions of water or oil surfaces.

In order to apply the law accurately to the experiment, one needs to express the precise volume occupied by the liquid inside the tube, and the volume of water displaced outside the tube. For that purpose, more detailed annotation must be introduced on the dimensional features of the test tube as shown in Fig. 1 – 7.

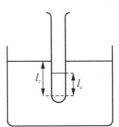

external
cross-section → S_z

internal
cross-section → S_c

arbitrary fixed
point A

section of the tube
where the cross-section
is assumed uniform

Fig. 1 – 6. Experimental configuration and experimental quantities to be measured.

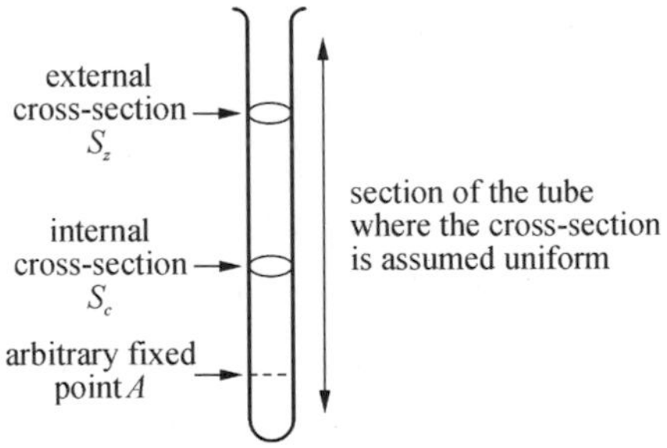

Fig. 1 – 7. Specification of test tube.

Theoretical formulation

The complete listing of notations to be used in the theoretical formulation along with their corresponding definitions are given below

S_c = internal cross-section of the tube above point A;

S_z = external cross-section of the tube above point A;

V_0 = internal volume of the test tube below point A;

V_e = external volume of the test tube below point A

= V_0 + volume of the glass below point A;

l_z = distance between point A and the water surface outside the tube;

l_c = distance between point A and the liquid surface inside the tube;

ρ_c = density of the liquid inside the tube

$\quad = \rho_w$ for water

$\quad = \rho_o$ for oil;

M = mass of the empty test tube.

At equilibrium, the buoyancy or the Archimedes force F_A is equal to the total weight W of the test tube including the liquid inside it. Referring to Figs. 1 – 6 and 1 – 7 as well as the notations listed above, we are led to the following expressions:

$$F_A = (V_e + S_z l_z)\rho_w g,$$
$$W = (M + V_0 \rho_c + S_c l_c \rho_c)g.$$

The equilibrium condition specified by $F_A = W$ implies

$$(V_e + S_z l_z)\rho_w = M + (V_0 + S_c l_c)\rho_c.$$

This equation can be put into the form:

$$l_z = C + D l_c,$$

where

$$D = \frac{\rho_c S_c}{\rho_w S_z},$$

$$C = \frac{M + V_0 \rho_c - V_e \rho_w}{\rho_w S_z}.$$

Since the coefficient D does not depend on the zero point of l_z and l_c, the reference point A in this experiment can be chosen at some convenient point on the tube within its length of uniform cross-section as implied by the above formulation.

Measurements

In the first part of the experiment, water is used as the liquid filling the test tube to various levels corresponding to different sets of values for the pair l_z and l_c. Plotting l_z as a function of l_c on the graph paper leads to the determination of D_1,

$$D_1 = \frac{S_c}{S_z},$$

since $\rho_c = \rho_w$ in this case.

The same measurements are repeated in the second part, replacing water with oil for the liquid inside the tube. The result is given by

$$D_2 = \frac{\rho_o S_c}{\rho_w S_z}.$$

Equation $\dfrac{S_c}{S_z}$ from the two equations results in the relation

$$\rho_o = \left(\frac{D_2}{D_1}\right)\rho_w.$$

Experimental result

The experimental results consist of two parts. The results shown in Tables 1 and 2 were obtained by using water and the oil as the filling liquids respectively.

Table 1.   Data from experiment 1 (water).

distance from the bottom	l_c (cm)	l_z (cm)
2.5	3.7	11.7
2.0	4.5	12.3
1.7	4.9	12.6
1.4	5.2	12.9
1.3	5.3	13.0
1.0	5.7	13.3

Table 2.   Data from experiment 2 (oil).

distance from the bottom	l_c (cm)	l_z (cm)
1.8	5.7	12.5
1.7	6.0	12.6
1.5	6.0	12.8
1.3	6.4	13.0
1.0	6.8	13.3
0.8	7.2	13.5

The value of D_1 determined from the slope of the plot in Fig. $1-8$ is

$$D_1 = 0.8091.$$

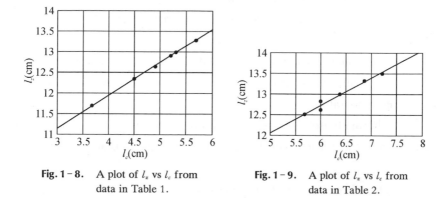

Fig. 1 - 8. A plot of l_z vs l_c from data in Table 1.

Fig. 1 - 9. A plot of l_z vs l_c from data in Table 2.

The value of D_2 determined from the slope of the plot in Fig. $1-9$ is

$$D_2 = 0.6865.$$

The final result for ρ_o & $\Delta\rho_o$ are

$$\rho_o = 0.8484 \text{ g/cm}^3,$$
$$\Delta\rho_o = 0.04\%.$$

Remarks

(1) For typical test tube, the ratio $\dfrac{S_c}{S_z}$ is about 0.8 instead of 1.

(2) All water and liquid surface positions to be measured must lie within the length of the tube with uniform cross-section.

(3) For the determination of D_1 and D_2, one should try to get more than 5 data points and draw the line with the best fit.

(4) The test tube should be dried before measurement with the oil.

(5) The most crucial problem in this experiment is how to get enough data (more than two data points) with the narrowly limited range of $2-2.5$ cm available for l_z variation.

Problem 2

Determination of the Stefan - Boltzmann Constant

Listed below are the only apparatus and materials available for your experiment:

(1) DC power supply;

(2) Heater mounted on a ceramic base;

(3) Digital voltmeter (labeled V) and ammeter (labeled A);

(4) Caliper;

(5) Aluminum cylinder with polished surface and a hole to house the heater. The cylinder is fitted with a thermocouple (iron-constant) for measuring temperature;

(6) Thermally isolated vessel containing water and ice for maintaining the cold (reference) junction of the thermocouple at the constant temperature of 0°C;

(7) Digital mV-meter (labeled mV) to be connected with the thermocouple;

(8) A table listing the calibrated thermoelectric characteristics of the thermocouple for converting the mV readings into the corresponding temperatures;

(9) Electric cables;

(10) Candle and safety matches for blackening the cylinder.

A note on the theoretical principle:

The reflective radiation of power by an object with surface area S at absolute temperature T in equilibrium with its surrounding is given by the formula

$$P = e\sigma S(T^4 - T_0^4),$$

where σ is the Stefan – Boltzmann constant, T_0 is the absolute temperature of the surroundings, and $e = 1$ for an ideal blackbody while $e = 0$ for an ideal reflector.

The room temperature will be given.

In this experiment, you are to determine the Stefan – Boltazmann constant. Include the following in your report:

(a) The theoretical basis for the measurement;

(b) A description of the method and procedure of the experiment;

(c) The final value of the Stefan – Boltzmann constant σ;

(d) The error and their sources.

Warnings:

(1) Be careful in handling some of the elements during the experiment as they may become very hot (100°C) at some stage.

(2) Be sure that the power supply current for the heater never exceeds 2A at all stages of the experiment.

Solution

Theoretical consideration

According to the theory of electromagnetic radiation of solids, the polished aluminium cylinder which can be regarded as an ideal reflector, does not absorb nor emit any radiation. On the other hand, the same cylinder covered by a thin layer of candle's soot is assumed to behave as an ideal black body which is a perfect absorber and emitter of thermal radiation.

Therefore, the hot polished cylinder is expected to lose its thermal energy by means of non-radiative mechanism, such as thermal conductivity and convection of surrounding air. In contrast, the hot blackened cylinder will lose its thermal energy by an additional process of thermal radiation according to Stefan – Boltzmann law.

Based on the different physical processes described above, 3 different methods of experiment can be formulated as follows:

(1) Method of constant temperature

Assume that the cylinder is heated to the same temperature T when it is unblackened (polished) and when it is blackened by the soot. The difference in the measured electric power needed to reach that same equilibrium temperature must be equal to power loss to radiative process. In other words,

$$P_r(T) = P_t(T) - P_n(T),$$

where

$P_r(T)$ = power loss of the blackened cylinder due to thermal radiation,

$P_t(T)$ = total power loss of the blackened cylinder at T,

$P_n(T)$ = power loss of the polished cylinder at T due to nonradiative processes.

Assuming the same $P_n(T)$ in both cases (polished and blackened), one obtains

$$\sigma = \frac{P_t(T) - P_n(T)}{S(T^4 - T_0^4)},$$

where T_0 is the surrounding (or room) temperature.

Alternatively, although less accurately, other methods may also be formulated by explicitly assuming that P_n is proportional to $(T - T_0)$, namely

$$P_n(T) = k(T - T_0),$$

where k is a constant independent of T. On the basis of this relation, one can formulate the following two methods for the determination of σ.

(2) Method of constant power

In this method, the power of heating P is kept the same in both cases. Let the temperatures reached in equilibrium for the polished and blackened cylinder be denoted by T_P and T_b respectively. Then,

$$P = k(T_P - T_0),$$
$$P = k(T_b - T_0) + P_r(T_b).$$

Eliminating k yields

$$P_r(T_b) = \frac{(T_P - T_b)}{(T_P - T_0)}.$$

Equating this to the radiative power expression of the Stefan - Boltzmann law, we obtain

$$\sigma = \frac{(T_P - T_b)P}{S(T_b^4 - T_0^4)(T_P - T_0)}.$$

(3) Method of two temperatures

In this case, the measurements are performed for the blackened cylinder only, but at two equlibrium temperatures T_1 and T_2. Let the heating powers required to reach T_1 and T_2 be P_1 and P_2 respectively. Then we have

$$P_1 = k(T_1 - T_0) + \sigma S(T_1^4 - T_0^4),$$

$$P_2 = k(T_2 - T_0) + \sigma S(T_2^4 - T_0^4).$$

Again, eliminating k from the two equations above leads directly to the following expression

$$\sigma = \frac{(T_2 - T_0)P_1 - (T_1 - T_0)P_2}{S[(T_1^4 - T_0^4)(T_2 - T_0) - (T_2^4 - T_0^4)(T_1 - T_0)]}.$$

Remarks

(1) The formulation of the first experimental method requires the insurance of the same T in both cases. Since P is proportional to T^4, a small difference in T determined in two cases will result in great error. It is, however, not easy to satisfy the requirement mentioned above. One way of overcoming this difficulty is to measure the power P_t for heating up the blackened cylinder at two temperatures in the vicinity of the temperature reached by the unblackened cylinder, and interpolate the value of P_t at the right T.

(2) It is also worth noting that due to the sensitivity of the measurement, a slight change in the surrounding of the cylinder is likely to affect the result significantly. The environment must therefore be kept constant during the experiment.

Experimental configuration

The experimental set-up is described in Fig. 1 – 10. The heater is mounted on a porcelain base, and it is connected with a power supply and the measuring meters. The heater is entirely enclosed by the hollow cylinder which sits also on the same porcelain plate during the measurement. The thermocouple is permanently attached to the cylinder and connected to an mV-meter for the determination of the temperature by using a table listing the characteristics of the thermocouple. The size of the cylinder is 60 mm by length and 12. 5 mm by its external diameter, leading to a surface area of $S = 24.\ 8\ \text{cm}^2$. The wall of the cylinder is about 1 mm thick and the thickness of its base is about 3 mm. All electrical measuring meters are digital instruments.

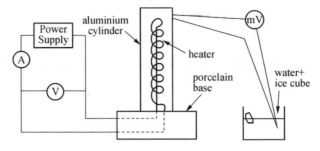

Fig. 1 – 10.   Experimental set-up.

The power supplied to the heater must be measured separately instead of being read off the power supply display panel, because the resistance of the heater varies somewhat with temperature. The reading of V and I should be done at thermal equilibrium between the cylinder and its surrounding, which will be reached in about 25 – 30 minutes. In order to avoid undesirable effects from the surrounding, the whole system should be kept at a distance from other objects in the laboratory.

Results of measurement

In a set of experiment performed at room temperature of 298.8 K, the results obtained are represented by the sample data given in Table 1.

Table 1.   The values of σ found in a set of three measurements.

Code name for the data	Surface condition during measurement	Data		
		V	A	$T(\text{K})$
a	polished	9.8	1.50	485.5
b	blackened	9.8	1.50	433.5
c	blackened	11.9	1.82	485.5

Table 2.   Results of σ obtained by three different methods.

Method used	Data used	Experimental result $\sigma_{\text{ex}}(\text{Wm}^{-2}\text{K}^{-4})$	$\sigma_{\text{ex}}/\sigma$
constant T	$a+c$	5.945×10^{-8}	1.05
constant P	$a+b$	6.087×10^{-8}	1.07
two T	$b+c$	5.386×10^{-8}	0.95

Discussion

While the last two methods are supposed to be less accurate than the first one, this is not always confirmed by the experimental results, as the control of experimental condition is not perfect. The major factors affecting the accuracies of the experimental results are enumerated and discussed as follows:

(1) The cylinder is not necessarily an ideal reflector when it surface is polished, nor is it an ideal black body when its surface is blackened by the candle's soot. In other words, the absorption coefficient is likely to be larger than 0 in the first case, and less than 1 for the second case. Both of these effects leads to lower value of σ.

(2) The heat losses via the porcelain base are out of control. Neglecting these losses will lead to deviation of σ from its real value.

(3) The resistivities of the connecting cables have been neglected also, leading to larger value of σ.

(4) The assumption of equal non-radiative loss for the case with polished and blackened surfaces is at best an approximation. For instance, the difference between thermal conductivity of the soot and that of aluminium is neglected in this experiment, leading to lower value of σ. The equality will also be violated due to uncontrollable heat losses via the porcelain base.

(5) The influences of air convection in the surrounding of the cylinder due to motions of the experimentator and other objects are also possible sources of errors.

Minutes of the Second Asian Physics Olympiad

Chinese Taipei, April 22 - May 1,2001

1. The following 12 countries were present at the 2nd Asian Physics Olympiad:

Australia, India, Indonesia, Israel, Jordon, Kazakhstan, Malaysia, Mongolia, Singapore, Thailand, Chinese Taipei, Vietnam.

Two countries were represented with observers:

Japan and Qatar.

2. Result of marking the papers by the organizers were presented: The best score (34.50 points) was achieved by Tsai Hsin Yu from Chinese Taipei (Absolute winner of the 2nd APhO). The second and third were Wang Chia Chun (Chinese Taipei) 31.00 points and Bui Le Na (Vietnam) 30.80 points. The following limits for awarding the medals and the honorable mention were established according to the statutes:

Gold Medal:	28 points,
Silver Medal:	25 points,
Bronze Medal:	20 points,
Honorable Mention:	16 points.

According to the above limits 7 Gold medals, 5 Silver medals, 11 Bronze medals and 16 honorable mentions were distributed to all delegations.

3. In addition to the regular prizes a number of special prizes were awarded:

● For the best score in Theory: Chen Wei Yin (Chinese Taipei)

● For the best score in Experiment: Tsai Hsin Yu (Chinese Taipei)

● For the most creative solution in Experiment: Bui Le Na (Vietnam)

● For the most creative solution in Theory: Rezy Pradipta (Indonesia)

● For the best female participant: Tsai Hsin Yu (Chinese Taipei)

● For the best participant among the new participating countries: Obed Tsur (Israel)

4. The International Board has unanimously elected Prof. Lin Ming Juey to post of secretary of the Asian Physics Olympiad for the next four years.

5. President of the APhO's distributed a list of the organizers of the competition in the future. It is:

● 2002 - Singapore (invitation disseminated during the 2nd APhO)

- 2003 – Thailand (confirmed orally)
- 2004 – Vietnam (confirmed)

6. The International Board expressed deep thanks to Prof. Lin Ming Juey and his collaborators for excellent conducting of the competition. The International Board highlighted all the difficulties occurring when the second event is organized and congratulated the organizers for successfully solving all of them.

7. The International Board agreed to put the best or the most creative solutions in APhO proceeding without asking the student's permission.

8. The International Board agreed to include Australia as a member of Asian Physics Olympiad and keeping the name of the competition as it is.

9. The International Board has unanimously accepted Prof. Lin Ming Juey to represent APhO in the World Federation of Physics Competition.

10. Acting on behalf the organizers of the next Asian Physics Olympiad a Singapore delegation announced the 3rd Asian Physics Olympiad will be organized in Singapore on May 6 – May 15, 2002 and codially invited all the participant countries to attend the competition.

Taipei April 30, 2001

Ming Juey, Lin Ph. D
Executive Director of the Organizing
Committee of the 2nd APhO
Secretary of the APhO

Yohanes Surya Ph. D
President of the APhO

Theoretical Competition

April 24, 2001   Time available: 5 hours

Problem 1
When will the Moon become a Synchronous Satellite?

The period of rotation of the Moon about its axis is currently the same as its period of revolution about the Earth so that the same side of the Moon always faces the Earth. The equality of these two periods presumably came about because of actions of tidal forces over the long history of the Earth-Moon system.

However, the period of rotation of the Earth about its axis is currently shorter than the period of revolution of the Moon. As a result, lunar tidal forces continue to act in a way that tends to slow down the rotational speed of the Earth and drive the Moon itself further away from the Earth.

In this question, we are interested in obtaining an estimate of how much more time it will take for the rotational period of the Earth to become equal to the period of revolution of the Moon. The Moon will then become a synchronous satellite, appearing as a fixed object in the sky and visible only to those observers on the side of the Earth facing the Moon. We also want to find out how long it will take for the Earth to complete one rotation when the said two periods are equal.

Two right-handed rectangular coordinate systems are adopted as reference frames. The third coordinate axes of these two systems are parallel to each other and normal to the orbital plane of the Moon.

(I) The first frame, called the *CM* frame, is an inertial frame with its origin located at the center of mass C of the Earth-Moon system.

(II) The second frame, called the *xyz* frame, has its origin fixed at the center O of the Earth. Its z-axis coincides with the axis of rotation of the Earth. Its x-axis is along the line connecting the centers of the Moon and the Earth, and points in the direction of the unit vector r as shown in Fig. 2 - 1.

The Moon remains always on the negative x-axis in this frame.

Note that distances in Fig. 2 – 1 are not drawn to scale. The curved arrows show the directions of the Earth's rotation and the Moon's revolution. The Earth-Moon distance is denoted by r.

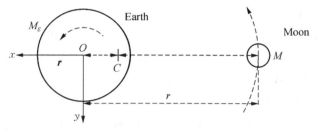

Fig. 2 – 1.

The following data are given:

(a) At present, the distance between the Moon and the Earth is $r_0 = 3.85 \times 10^8$ m and increases at a rate of 0.038 m per year.

(b) The period of revolution of the Moon is currently $T_M = 27.322$ days.

(c) The mass of the Moon is $M = 7.35 \times 10^{22}$ kg.

(d) The radius of the Moon is $R_M = 1.74 \times 10^6$ m.

(e) The period of rotation of the Earth is currently $T_E = 23.933$ hours.

(f) The mass of the Earth is $M_E = 5.97 \times 10^{24}$ kg.

(g) The radius of the Earth is $R_E = 6.37 \times 10^6$ m.

(h) The universal gravitational constant is $G = 6.672\ 59 \times 10^{-11}$ N $\cdot$ m^2/kg^2.

The following assumptions may be made when answering questions:

(i) The Earth-Moon system is isolated from the rest of the universe.

(ii) The orbit of the Moon about the Earth is circular.

(iii) The axis of rotation of the Earth is perpendicular to the orbital plane of the Moon.

(iv) If the Moon is absent and the Earth does not rotate, then the mass distribution of the Earth is spherically symmetric and the radius of the Earth is R_E.

(v) For the Earth or the Moon, the moment of inertia I about any axis passing through its center is that of a uniform sphere of mass M and radius R, i.e. $I = \dfrac{2MR^2}{5}$.

(vi) The water around the Earth is stationary in the xyz frame.

Answer the following questions:

(1) With respect to the center of mass C, what is the current value of the total angular momentum L of the Earth-Moon system?

(2) When the period of rotation of the Earth and the period of revolution of the Moon become equal, what is the duration of one rotation of the Earth? Denote the answer as T and express it in units of the present day. Only an approximate solution is required so that iterative methods may be used.

(3) Consider the Earth to be a rotating solid sphere covered with a surface layer of water and assume that, as the Moon moves around the Earth, the water layer is stationary in the xyz-frame. In one model, frictional forces between the rotating solid sphere and the water layer are taken into account. The faster spinning solid Earth is assumed to drag lunar tides along so that the line connecting the tidal bulges is at an angle δ with the x-axis, as shown in Fig. 2 – 2. Consequently, lunar tidal forces acting on the Earth will exert a torque Γ about O to slow down the rotation of the Earth.

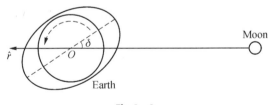

Fig. 2 – 2.

The angle δ is assumed to be constant and independent of the Earth-Moon distance r until it vanishes when the Moon's revolution is synchronous with the Earth's rotation so that frictional forces no longer exist. The torque Γ therefore scales with the Earth-Moon distance and is proportional to $1/r^6$.

According to this model, when will the rotation of the Earth and the revolution of the Moon have the same period? Denote the answer as t_f and express it in units of the present year.

The following mathematical formulae may be useful when answering questions:

(M1) For $0 \leqslant s < r$ and $x = s\cos\theta$:

$$\frac{1}{\sqrt{r^2 + s^2 + 2rx}} \approx \left(\frac{1}{r} - \frac{x}{r^2} + \frac{3x^2 - s^2}{2r^3} + \cdots \right).$$

(M2) If $a \neq 0$ and $\dfrac{d\omega}{dt} = b\omega^{1-a}$, then $\omega^a(t') - \omega^a(t) = (t' - t)ab$.

Solution

(1) The total angular momentum $L = Lz$ of the Earth-Moon system with respect to C can be calculated as follows.

Since all angular momenta are along the z-direction, only the z-component of each angular momentum have to be calculated. The distance between the center of mass C and the center of the Earth O is

$$r_{CM} = \frac{Mr_0}{M + M_E} = \frac{3.85 \times 10^8}{1 + (597/7.35)} = 4.68 \times 10^6 \text{ m} = 0.735R_E.$$

The angular speed of the Moon's revolution is

$$\omega_0 = \frac{2\pi}{27.322 \times 86\,400} = 2.6617 \times 10^{-6} \text{ rad/s}. \tag{1a}$$

The orbital angular momentum of the Moon about C is

$$\begin{aligned}
L_M &= M(r_0 - r_{CM})^2 \omega_0 \\
&= 7.35 \times (385 - 4.68)^2 \times 2.6617 \times 10^{28} \\
&= 2.83 \times 10^{34} \text{ kg} \cdot \text{m}^2/\text{s}.
\end{aligned}$$

The angular speed of the Moon's spinning or rotational motion is

$$\Omega_M = \omega_0 = 2.6617 \times 10^{-6} \text{ rad/s}.$$

The spin angular momentum of the Moon is

$$S_M = \frac{2}{5}MR_M^2\Omega_M = \frac{2}{5} \times 7.35 \times (1.74)^2 \times 2.6617 \times 10^{28}$$

$$= 2.37 \times 10^{29} \text{ kg} \cdot \text{m}^2/\text{s} = 8.40 \times 10^{-6}L_M.$$

This is much smaller than the Moon's orbital angular momentum and can therefore be neglected.

The orbital angular momentum of the Earth about C is

$$L_E = M_E r_{CM}^2 \omega_0 = \frac{M}{M_E}L_M$$

$$= \frac{7.35}{597} \times 2.83 \times 10^{34}$$

$$= 3.48 \times 10^{32} \text{ kg} \cdot \text{m}^2/\text{s}.$$

The angular speed of the Earth's spinning motion is

$$\Omega_E = \frac{2\pi}{23.933 \times 3600} = 7.2926 \times 10^{-5} \text{ rad/s}.$$

The moment of inertia of the Earth about its axis of rotation is

$$I = \frac{2}{5}M_E R_E^2 = 0.4 \times 5.97 \times (6.37)^2 \times 10^{36}$$

$$= 9.69 \times 10^{37} \text{ kg} \cdot \text{m}^2/\text{s}. \tag{1b}$$

The spin angular momentum of the Earth is

$$S_E = \frac{2}{5}M_E R_E^2 \Omega_E = 7.07 \times 10^{33} \text{ kg} \cdot \text{m}^2/\text{s} = 20.3L_E.$$

Thus, the total angular momentum of the Earth-Moon system L is given by

$$L = (L_M + L_E + S_E + S_M)$$

$$= (2.83 + 0.0348 + 0.707 + 0.000\ 023\ 7) \times 10^{34}$$

$$= 3.57 \times 10^{34} \text{ kg} \cdot \text{m}^2/\text{s}. \tag{2}$$

Note that $L \approx (L_M + L_E + S_E)$.

(2) According to Newton's form for Kepler's third law of planetary motions, the angular speed ω of the revolution of the Moon about the Earth is related to the Earth-Moon distance r by

$$\omega^2 r^3 = G(M_E + M). \tag{3}$$

Therefore, the orbital angular momentum of the Earth-Moon system with respect to C is

$$L = \left(\frac{M_E M}{M + M_E}\right) r^2 \omega = MM_E \left(\frac{G^2}{\omega(M + M_E)}\right)^{\frac{1}{3}}. \tag{4}$$

$\left(\text{Note: } L_M = M\left(\frac{M_E r}{M + M_E}\right)^2 \omega, \ L_E = M_E\left(\frac{Mr}{M + M_E}\right)^2 \omega, \text{ so that } L = L_E + L_M.\right)$

When the angular speed of the Earth's rotation is equal to the angular speed ω of the orbiting Moon, the total angular momentum of the Earth-Moon system is, with the spin angular momentum of the Moon neglected, given by

$$
\begin{aligned}
(L_M + L_E + S_E) &= MM_E \left\{\frac{G^2}{(M + M_E)\omega}\right\}^{\frac{1}{3}} + \frac{2}{5} M_E R_E^2 \omega \\
&= 7.35 \times 5.97 \times \left\{\frac{66.726 \times 66.726}{(5.97 + 0.0735)\omega}\right\}^{\frac{1}{3}} \\
&\quad \times 10^{30} + 9.69 \times 10^{37} \omega \\
&= 3.96 \times 10^{32} \omega^{-1/3} + 9.69 \times 10^{37} \omega \\
&= 3.57 \times 10^{34} \text{ kg} \cdot \text{m}^2/\text{s}.
\end{aligned} \tag{5a}
$$

The last equality follows from conservation of total angular momentum and Eq. (2). For an initial estimate of ω, the spin angular momentum of the Earth may be neglected in Eq. (5a) to give

$$\omega \approx \omega_1 = \left(\frac{3.96}{357}\right)^3 = 1.36 \times 10^{-6} \text{ rad/s. } (\textit{first iteration})$$

An improved estimate may be obtained by using the above estimated value ω_1 to compute the spin angular momentum of the Earth and use Eq. (5a) again to solve for ω. The result is

$$\omega \approx \omega_f = \left(\frac{3.96}{356}\right)^3 = 1.38 \times 10^{-6} \text{ rad/s. } (\textit{second iteration}) \tag{5b}$$

Further iterations of the same procedure lead to the same value just given. Thus, the period of rotation of the Earth will be

$$T_f = \frac{2\pi}{\omega_f} = \frac{6.2832}{1.38 \times 10^{-6} \times 86\,400} = 52.7 \text{ days.}$$

(3) Since the total torque Γ is proportional to $1/r^6$, we conclude

$$r^6 \Gamma = \text{constant.} \tag{6}$$

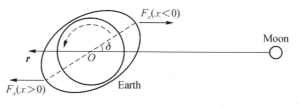

Fig. 2 - 3.

Let the current values of r and Γ be, respectively, r_0 and Γ_0. From Eq. (6), we then have

$$\Gamma = \left(\frac{r_0}{r}\right)^6 \Gamma_0. \tag{7}$$

The torque Γ is equal to the *rate of change* of spin angular momentum $I\Omega$ of the Earth so that

$$I\frac{d\Omega}{dt} = \Gamma. \tag{8}$$

By Newton's law of action and reaction or by the law of conservation of the total angular momentum, $-\Gamma$ is equal to the *rate of change* of the total orbital angular momentum L of the *Earth-Moon system* so that

$$\frac{dL}{dt} = -\Gamma. \tag{9}$$

But according to Eq. (3), we have

$$\omega^2 r^3 = G(M_E + M),$$

and Eq. (4) may be written as

$$L = \left(\frac{MM_E}{M_E + M}\right)\omega\, r^2 = MM_E \left(\frac{G}{M_E + M}\right)^{\frac{1}{2}} r^{\frac{1}{2}}$$

$$= MM_E \left(\frac{G^2}{M_E + M}\right)^{\frac{1}{3}} \omega^{-\frac{1}{3}} \tag{10}$$

This implies

$$\frac{dL}{dt} = MM_E \left(\frac{G}{M_E + M}\right)^{\frac{1}{2}} \frac{1}{2r^{\frac{1}{2}}} \frac{dr}{dt}$$

$$= -\frac{1}{3} MM_E \left(\frac{G^2}{M_E + M}\right)^{\frac{1}{3}} \frac{1}{\omega^{\frac{4}{3}}} \frac{d\omega}{dt} = -\Gamma. \tag{11}$$

The value of Γ_0 can be determined from Eq. (11) as follows:

$$-\Gamma_0 = \left(\frac{dL}{dt}\right)_0 = \frac{1}{2} MM_E \sqrt{\frac{G}{(M_E + M)r_0}} \left(\frac{dr}{dt}\right)_0$$

$$= \frac{1}{2} \times 7.35 \times 5.97 \times 10^{46} \cdot \sqrt{\frac{66.7 \times 10^{-44}}{(0.0735 + 5.97) \times (3.85)}}$$

$$\cdot \frac{3.8 \times 10^{-8}}{3.65 \times 8.64}$$

$$= 4.5 \times 10^{16} \text{ N} \cdot \text{m.} \tag{12}$$

Starting with Eq. (11) in the following form

$$\frac{dL}{dt} = -\frac{1}{3} MM_E \left(\frac{G^2}{(M_E + M)}\right)^{\frac{1}{3}} \frac{1}{\omega^{\frac{4}{3}}} \frac{d\omega}{dt} = -\left(\frac{r_0}{r}\right)^6 \Gamma_0,$$

we may use Eq. (3) to express r in terms of ω and obtain the following equations:

$$\frac{1}{3} MM_E \left(\frac{G^2}{(M_E + M)}\right)^{\frac{1}{3}} \left(\frac{d\omega}{dt}\right) = \frac{(r_0)^6 \Gamma_0}{[G(M_E + M)]^2} \omega^{\frac{16}{3}},$$

$$\frac{d\omega}{dt} = \left\{\frac{3(r_0)^6 \Gamma_0}{GM_E M [G(M_E + M)]^{\frac{1}{3}}}\right\} \omega^{\frac{16}{3}} = b\omega^{\frac{16}{3}},$$

where the constant b stands for the expression in the square brackets. The last equation leads to the solution

$$(\omega_f)^{-\frac{13}{3}} - (\omega_0)^{-\frac{13}{3}} = \frac{-13b}{3} (t_f - 0),$$

where t_f is the length of time needed for the angular speed of the rotation of the Earth to be equal to that of the Moon's revolution about the Earth.

Using the values of ω_f and ω_0 obtained in Eqs. (1a) and (5b) and the value of Γ_0 in Eq. (12), we have

$$\frac{-3}{13b} = \frac{GM_E M [G(M_E + M)]^{\frac{5}{3}}}{13(r_0)^6(-\Gamma_0)} = 3.4 \times 10^{-8},$$

and accordingly

$$t_f = \frac{-3}{13b}(\omega_f^{-\frac{13}{3}} - \omega_0^{-\frac{13}{3}})$$

$$= 3.4 \times [(1.38)^{-\frac{13}{3}} - (2.6617)^{-\frac{13}{3}}] \times 10^{18}$$

$$= 3.4 \times 10^{18} \times (0.248 - 0.014\ 376)$$

$$= 7.9 \times 10^{17} \text{ s}$$

$$= 2.5 \times 10^{10} \text{ years.}$$

Problem 2
Motion of an Electric Dipole in a Magnetic Field

In the presence of a constant uniform magnetic field B, the translational motion of a system of electric charges is coupled to its rotational motion. As a result, conservation laws relating to the momentum and the component of the angular momentum along the direction of B must be expressed in modified forms different from the usual. This is studied in this problem by considering motions of an electric dipole formed by two particles of the same mass m, but opposite charges q ($q > 0$) and $-q$. A rigid insulating thin rod of length l with negligible mass connects the two particles. Let $l = r_1 - r_2$ with r_1 and r_2 representing position vectors of the particle with charge q and $-q$, respectively. Denote by ω the angular velocity of the rotational motion of the dipole about its center of mass. Denote by r_{CM} and v_{CM} the position and the velocity vectors, respectively, of its center of mass. Relativistic effects, radiation of electromagnetic waves, and rotation of the dipole about the line connecting the particles may be neglected.

Note that the magnetic force acting on a particle of charge q and velocity v is $qv \times B$, where the cross product $A_1 \times A_2$ of two vectors A_1 and A_2 is defined, in terms of x, y and z components of the vectors in a rectangular coordinate system, by

$$(A_1 \times A_2)_x = (A_1)_y(A_2)_z - (A_1)_z(A_2)_y,$$
$$(A_1 \times A_2)_y = (A_1)_z(A_2)_x - (A_1)_x(A_2)_z,$$
$$(A_1 \times A_2)_z = (A_1)_x(A_2)_y - (A_1)_y(A_2)_x.$$

(1) Conservation laws

(a) For the dipole, compute the total force on it and the total torque on it with respect to the center of mass and write down equations of motion for its center of mass and for rotation about its center of mass.

(b) From the equation of motion for the center of mass, deduce the modified form of the conservation law for the total momentum with P denoting the conserved quantity. Write down an expression in terms of v_{CM} and ω for the conserved energy E.

(c) The angular momentum of the dipole consists of two parts. One part is due to the motion of its center of mass and the other is due to rotation about the center of mass. From the modified form of the conservation law for the total momentum and the equation of motion for rotation about the center of mass, prove that the quantity J as defined by

$$J = (r_{CM} \times P + I\omega) \cdot B$$

is conserved, where I is the moment of inertia about an axis passing through the center of mass in a direction perpendicular to l and B is the unit vector along the magnetic field.

Note that

$$A_1 \times A_2 = -A_2 \times A_1,$$
$$A_1 \cdot (A_2 \times A_3) = (A_1 \times A_2) \cdot A_3,$$
$$A_1 \times (A_2 \times A_3) = (A_1 \cdot A_3)A_2 - (A_1 \cdot A_2)A_3,$$

for any three vectors A_1, A_2 and A_3. Repeated application of the first two formulas above may be useful in deriving the conservation laws in question.

(2) Motion in a plane perpendicular to B

Let the constant magnetic field be in the z-direction so that $B = B z$ with z denoting the unit vector in the z-direction. In the following, we assume the dipole moves only in the $z = 0$ plane so that $\omega = \omega z$. Suppose initially the center of mass of the dipole is at rest at the origin such that l points in the +

x-direction and the initial angular velocity of the dipole is ω_0 z.

(a) If the magnitude of ω_0 is smaller than a critical value ω_c, the dipole will not make a full turn with respect to its center of mass. Find ω_c.

(b) For a general $\omega_0 > 0$, what is the maximum distance d_m in the x-direction that the center of mass can reach?

(c) What is the tension on the rod? Express it as a function of the angular velocity ω.

Solution

(1) Conservation laws

(1a) Denote the velocities of the two particles by v_1 and v_2. We have

$$r_{CM} = \frac{1}{2}(r_1 + r_2), \ v_{CM} = \frac{1}{2}(v_1 + v_2), \ l = r_1 - r_2, \ \dot{l} = v_1 - v_2.$$

To compute the total force F on the dipole, only external forces due to the magnetic field B have to be considered. Hence one obtains

$$F = F_1 + F_2 = q(v_1 \times B) + (-q)(v_2 \times B)$$
$$= q(v_1 - v_2) \times B$$
$$= q\dot{l} \times B.$$

The equation of motion for the center of mass is therefore

$$M\dot{v}_{CM} = q\dot{l} \times B = q(v_1 - v_2) \times B \quad (M = 2m). \tag{1}$$

Similarly, the total torque about the center of mass is given by

$$\tau = \left(\frac{l}{2}\right) \times (qv_1 \times B) + \left(\frac{-l}{2}\right) \times (-qv_2 \times B)$$
$$= ql \times \left(\frac{v_1 + v_2}{2} \times B\right)$$
$$= ql \times (v_{CM} \times B).$$

Since the dipole does not rotate about the line connecting the particles, the angular momentum of the dipole about its center of mass is $L = I\omega$, where I denotes the moment of inertia about an axis passing through the center of mass in a direction perpendicular to l and is given by

$$I = m\left(\frac{l}{2}\right)^2 + m\left(\frac{l}{2}\right)^2 = \frac{1}{2}ml^2. \tag{2}$$

The equation of motion for rotation about the center of mass is then given by

$$\frac{\mathrm{d}\boldsymbol{L}}{\mathrm{d}t} = I\dot{\boldsymbol{\omega}} = \boldsymbol{\tau} = q\boldsymbol{l} \times (\boldsymbol{v}_{CM} \times \boldsymbol{B}). \tag{3}$$

(1b) From Eq. (1), we obtain the conservation law for the momentum:

$$\dot{\boldsymbol{P}} = 0, \ \boldsymbol{P} = M\boldsymbol{v}_{CM} - q\boldsymbol{l} \times \boldsymbol{B}. \tag{4}$$

The relative velocity of the two particles may be written as $\dot{\boldsymbol{l}} = (\boldsymbol{v}_1 - \boldsymbol{v}_2) = \boldsymbol{\omega} \times \boldsymbol{l}$.

Therefore from Eqs. (1) and (3), one obtains

$$\boldsymbol{v}_{CM} \cdot M\dot{\boldsymbol{v}}_{CM} + \boldsymbol{\omega} \cdot I\dot{\boldsymbol{\omega}} = q\boldsymbol{v}_{CM} \cdot \dot{\boldsymbol{l}} \times \boldsymbol{B} + q\boldsymbol{\omega} \cdot \boldsymbol{l} \times (\boldsymbol{v}_{CM} \times \boldsymbol{B})$$
$$= -q\dot{\boldsymbol{l}} \cdot \boldsymbol{v}_{CM} \times \boldsymbol{B} + q(\boldsymbol{\omega} \times \boldsymbol{l}) \cdot (\boldsymbol{v}_{CM} \times \boldsymbol{B})$$
$$= 0.$$

The left-hand side may be rewritten as

$$\boldsymbol{v}_{CM} \cdot M\dot{\boldsymbol{v}}_{CM} + \boldsymbol{\omega} \cdot I\dot{\boldsymbol{\omega}} = \frac{1}{2}\frac{\mathrm{d}}{\mathrm{d}t}(M\boldsymbol{v}_{CM} \cdot \boldsymbol{v}_{CM} + I\boldsymbol{\omega} \cdot \boldsymbol{\omega})$$
$$= \frac{1}{2}\frac{\mathrm{d}}{\mathrm{d}t}(Mv_{CM}^2 + I\omega^2).$$

From the last two equations, one obtains the conservation law for the energy as

$$\dot{E} = 0, \ E = \frac{1}{2}Mv_{CM}^2 + \frac{1}{2}I\omega^2. \tag{5}$$

(1c) Using Eqs. (3) and (4) and noting that $\boldsymbol{P}$ and $\boldsymbol{B}$ are both constant, one obtains

$$\frac{\mathrm{d}}{\mathrm{d}t}(I\boldsymbol{\omega} \cdot \boldsymbol{B}) = I\dot{\boldsymbol{\omega}} \cdot \boldsymbol{B} = \boldsymbol{B} \cdot I\dot{\boldsymbol{\omega}}$$
$$= q\boldsymbol{B} \cdot \boldsymbol{l} \times (\boldsymbol{v}_{CM} \times \boldsymbol{B}) = q(\boldsymbol{B} \times \boldsymbol{l}) \cdot (\boldsymbol{v}_{CM} \times \boldsymbol{B})$$
$$= (\boldsymbol{P} - M\boldsymbol{v}_{CM}) \cdot (\boldsymbol{v}_{CM} \times \boldsymbol{B})$$
$$= \boldsymbol{P} \cdot (\boldsymbol{v}_{CM} \times \boldsymbol{B}) = (\boldsymbol{P} \times \boldsymbol{v}_{CM}) \cdot \boldsymbol{B}$$

$$=-(v_{CM} \times P) \cdot B = -\frac{d}{dt}\{(r_{CM} \times P) \cdot B\}.$$

By transferring the last term to the left side of the equation, one obtains the conservation law

$$\dot{J} = 0 \qquad J = (r_{CM} \times P + I\omega) \cdot B \qquad (6)$$

for the component of the angular momentum along the direction of B.

(2) Motion in a plane perpendicular to B

(2a) Since the dipole remains in the $z = 0$ plane, we may write

$$l = l[\cos \varphi(t)\,x + \sin \varphi(t)\,y], \quad \varphi(0) = 0, \quad \dot{\varphi}(0) = \omega_0. \qquad (7)$$

The angular velocity may be written in terms of $\dot{\varphi}$ as

$$\omega = \omega z = \dot{\varphi} z. \qquad (8)$$

From Eq. (4), we have

$$Mv_{CM} = P + qlB(\sin \varphi\,x - \cos \varphi\,y). \qquad (9)$$

At $t = 0$, we have $v_{CM} = 0$ and $\varphi = 0$ so that the conserved quantity is given by

$$P = qlB\,y. \qquad (10)$$

Hence from Eqs. (9) and (10) we have

$$\dot{x}_{CM} = \left(\frac{qlB}{M}\right)\sin \varphi, \quad \dot{y}_{CM} = \left(\frac{qlB}{M}\right)(1 - \cos \varphi). \qquad (11)$$

From conservation of energy or Eq. (5), we have

$$\frac{1}{2}I\dot{\varphi}^2 + \frac{(qlB)^2}{M}(1 - \cos \varphi) = \frac{1}{2}I\omega_0^2$$

or equivalently

$$\dot{\varphi}^2 + \frac{1}{2}\omega_c^2(1 - \cos \varphi) = \omega_0^2, \qquad (12)$$

where

$$\omega_c = \sqrt{\frac{4(qlB)^2}{MI}} = \sqrt{\frac{4(qlB)^2}{2m\left(\frac{ml^2}{2}\right)}} = \frac{2qB}{m}. \qquad (13)$$

(Note that Eq. (12) has the same form as that obtained for a simple pendulum readily found by employing this analogy.)

In order to swinging in a vertical plane under gravity, answers for Problems (2a) and (2b) may therefore be able to make a full turn, $\dot{\varphi}$ can-not become zero so that from Eq. (12) we have

$$\omega_0^2 - \frac{1}{2}\omega_c^2(1 - \cos\varphi) = \dot{\varphi}^2 > 0.$$

When $\varphi = \pi$, $\dot{\varphi}^2$ reaches its minimum value of $\omega_0^2 - \omega_c^2$ and it follows

$$\omega_0^2 - \omega_c^2 > 0, \ |\omega_0| > \omega_c = \frac{2qB}{m}. \tag{14}$$

The critical value is therefore ω_c as given in Eq. (13).

(2b) Eq. (10) may be written as $\boldsymbol{P} = P\boldsymbol{y}$ so that $P = qlB > 0$. From Eq. (6), we have

$$x_{CM}P + I\omega = J. \tag{15}$$

At $t = 0$, $x_{CM} = 0$ and $\omega = \omega_0$. Thus $J = I\omega_0$ and Eq. (15) becomes

$$x_{CM}P = I(\omega_0 - \omega). \tag{16}$$

Since $\omega_0 > 0$ as stated in the problem and $\omega_0^2 \geqslant \dot{\varphi}^2 = \omega^2$ by Eq. (12), we have $\omega_0 \geqslant \omega$.

Eq. (16) then implies $x_{CM} \geqslant 0$. Thus x_{CM} reaches a maximum d_m when ω is at its minimum value.

If $\omega_0 < \omega_c$, the dipole undergoes oscillation about $\varphi = 0$ and the minimum value of ω is $-\omega_0$ so that

$$d_m = \frac{I(2\omega_0)}{P} = \left(\frac{m\omega_0}{qB}\right)l, \ \omega_0 < \omega_c. \tag{17}$$

If $\omega_0 > \omega_c$, then ω will never become zero and must remain positive from the beginning. The minimum value of ω is $\sqrt{\omega_0^2 - \omega_c^2}$ so that

$$d_m = \left(\frac{I}{P}\right)(\omega_0 - \sqrt{\omega_0^2 - \omega_c^2}) = \frac{m}{2qB}(\omega_0 - \sqrt{\omega_0^2 - \omega_c^2})l, \ \omega_0 > \omega_c. \tag{18}$$

If $\omega_0 = \omega_c$, then $\omega^2 = \omega_c^2 \dfrac{1 + \cos\varphi}{2} = \omega_c^2 \cos^2\dfrac{\varphi}{2}$ and $\dot{\varphi} = \omega_c \cos\dfrac{\varphi}{2}$. When φ

increases and becomes close to π, we let $\varphi = \pi - 2\varepsilon$. Then $\dot{\varepsilon} = -\omega_c \sin\frac{\varepsilon}{2} \approx$

$-\omega_c\frac{\varepsilon}{2}$ so that $\varepsilon \sim e^{-\frac{\omega_c t}{2}}$ and it will take $t \to \infty$ to make $\varepsilon \to 0$ or $\varphi \to \pi$. Hence

$\dot{\varphi} \geqslant 0$ and the minimum value of ω is zero and

$$d_\mathrm{m} = \left(\frac{I}{P}\right)\omega_0 = \left(\frac{m\omega_0}{2qB}\right)l, \ \omega_0 = \omega_c. \tag{19}$$

(2c) Let positive value of a force correspond to compression on the rod. The tension on the rod is equal to the sum of the following three parts:

(i) Coulomb force between particles $F_C = \dfrac{1}{4\pi\varepsilon_0}\dfrac{q^2}{l^2}$. $\qquad\qquad$ (20)

(ii) Centrifugal effect as the rod rotates $= -\dfrac{1}{2}m\omega^2 l$. $\qquad\quad$ (21)

(iii) l-component of the magnetic force as both particles undertake translational motion with their center of mass

$$= q\boldsymbol{v}_\mathrm{CM} \times \boldsymbol{B} \cdot (-\boldsymbol{l}) = q\boldsymbol{v}_\mathrm{CM} \cdot \boldsymbol{l} \times \boldsymbol{B}.$$

Squaring both sides of Eq. (4), we obtain

$$P^2 = (M v_\mathrm{CM})^2 - 2M\boldsymbol{v}_\mathrm{CM} \cdot q\boldsymbol{l} \times \boldsymbol{B} + (qlB)^2.$$

Using Eqs. (10) and (5) in the last equation gives

$$\frac{1}{2}M v_\mathrm{CM}^2 = q l \boldsymbol{v}_\mathrm{CM} \cdot \boldsymbol{l} \times \boldsymbol{B} = \frac{1}{2}I(\omega_0^2 - \omega^2). \tag{22}$$

Combining the three parts in (i) to (iii), we find the tension on the rod to be

$$T = \frac{1}{4\pi\varepsilon_0}\frac{q^2}{l^2} - \frac{1}{2}m l \omega^2 + \frac{1}{4}m l (\omega_0^2 - \omega^2). \tag{23}$$

Again, a positive value of T corresponds to compression on the rod.

An alternative way of solution is by using the equation of motion for the particle with charge q. Its acceleration is given by

$$\frac{d\boldsymbol{v}_1}{dt} = \frac{d}{dt}\left(\boldsymbol{v}_\mathrm{CM} + \frac{1}{2}\boldsymbol{l}\right) = \dot{\boldsymbol{v}}_\mathrm{CM} + \frac{1}{2}\frac{d}{dt}(\boldsymbol{\omega} \times \boldsymbol{l})$$

$$= \dot{\boldsymbol{v}}_\mathrm{CM} + \frac{1}{2}[(\dot{\boldsymbol{\omega}} \times \boldsymbol{l}) + (\boldsymbol{\omega} \times \dot{\boldsymbol{l}})],$$

where we have made use of the relation $\dot{l} = \omega \times l$. The total force on this particle is

$$F = (T - F_C)\, l + q v_1 \times B$$
$$= (T - F_C)\, l + q\left(v_{CM} + \frac{1}{2}\,\dot{l}\right) \times B.$$

According to Newton's second law of motion, we have

$$l \cdot m\,\frac{\mathrm{d}v_1}{\mathrm{d}t} = l \cdot F$$

or equivalently

$$l \cdot m\,\dot{v}_{CM} + \frac{m}{2}\, l \cdot (\omega \times \dot{l}) = T - F_C + q\, l \cdot v_{CM} \times B + \frac{q}{2}\, l \cdot \dot{l} \times B.$$

According to Eq. (1), the first and the last terms in the preceding equation are equal and we have therefore

$$T = F_C - q\, l \cdot v_{CM} \times B + \frac{m}{2}\, l \cdot (\omega \times \dot{l})$$

$$= F_C - q\, l \cdot v_{CM} \times B - \frac{m}{2}\,(\omega \times l) \cdot \dot{l}$$

$$= F_C - q\, l \cdot v_{CM} \times B - \frac{m}{2}\,(\omega \times l) \cdot (\omega \times l)$$

$$= F_C + q v_{CM} \cdot l \times B - \frac{1}{2}\,m l \omega^2.$$

This is the same answer as given in Eq. (23).

Problem 3
Thermal Vibrations of Surface Atoms

This question considers the thermal vibrations of surface atoms in an elemental metallic crystal with a face-centered cubic (*fcc*) lattice. For the crystal under consideration, the unit cubic cell of its *fcc* lattice has one atom at each corner and one atom at the center of each face of the cubic cell, as shown in Fig. 2 – 4. We use $(a, 0, 0)$, $(0, a, 0)$ and $(0, 0, a)$ to represent the locations of the three atoms on the x, y and z axes of the unit cell. The lattice constant a is equal to 3. 92 Å (i. e. the length of each side of the cube is 3. 92 Å).

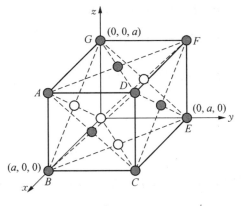

Fig. 2 – 4.

(1) The crystal is cut in such a way that the plane containing $ABCD$ becomes a boundary surface and is chosen for doing low-energy electron diffraction experiments. A collimated beam of electrons with kinetic energy of 64. 0 eV is incident on this surface plane at an incident angle ϕ_0 of 15. 0°. Note that ϕ_0 is the angle between the incident electron beam and the normal of the surface plane. The plane containing $\overline{AC}$ and the normal of the surface plane is the plane of incidence. For simplicity, we assume that all incident electrons are back scattered only by the surface atoms on the topmost layer.

(a) What is the wavelength of the matter waves of the incident electrons?

(b) If a detector is set up to detect electrons that do not leave the plane of incidence after being diffracted, at what angles with the normal of the surface will these diffracted electrons be observable?

(2) Assume that thermal vibrational motions of the surface atoms are simple harmonic. As the temperature rises, the amplitude of vibration increases. The average amplitude of vibration can be measured by means of low-energy electron diffraction. The intensity I of the diffracted beam is proportional to the number of scattered electrons per second. The relation between the intensity I and the displacement $u(t)$ of a surface atom is given

by

$$I = I_0 \exp\{- \langle [(\mathbf{K'} - \mathbf{K}) \cdot \mathbf{u}]^2 \rangle\}. \qquad (1)$$

In Eq. (1), I and I_0 are intensities at temperature T and absolute zero, respectively. $\mathbf{K}$ and $\mathbf{K'}$ are wave vectors of incident electrons and diffracted electrons, respectively. The angle brackets $\langle \rangle$ is used to denote average over time. Note that the relation between the wave vector $\mathbf{K}$ and the momentum $\mathbf{p}$ of a particle is $\mathbf{K} = \dfrac{2\pi \mathbf{p}}{h}$, where h is the Planck constant.

To measure vibration amplitudes of surface atoms of a metallic crystal, a collimated electron beam with kinetic energy of 64.0 eV is incident on a crystal surface at an incident angle of 15.0°. The detector is set up for measuring specularly reflected electrons. Only elastically scattered electrons are detected. A plot of $\ln \dfrac{I}{I_0}$ versus temperature T is shown in Fig. 2 - 5.

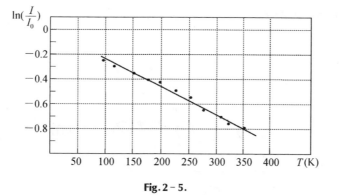

Fig. 2 - 5.

Assume the total energy of an atom vibrating in the direction of the surface normal x is given by $k_B T$, where k_B is the Boltzmann constant.

(a) Calculate the frequency of vibration in the direction of the surface normal for the surface atoms.

(b) Calculate the root-mean-square displacement, i. e. the value of $(\langle u_x^2 \rangle)^{\frac{1}{2}}$, in the direction of the surface normal for the surface atoms at 300 K.

The following data are given:

Atomic weight of the metal $M = 195.1$.

Boltzmann constant $k_B = 1.38 \times 10^{-23}$ J/K.

Mass of electron $= 9.11 \times 10^{-31}$ kg.

Charge of electron $= 1.60 \times 10^{-19}$ C.

Planck constant $h = 6.63 \times 10^{-34}$ J · s.

Solution

(1a) The wavelength of the incident electron is

$$\lambda = \frac{h}{p} = \frac{h}{\sqrt{2meV}}$$

$$= \frac{6.63 \times 10^{-34}}{\sqrt{2 \times 9.11 \times 10^{-31} \times 1.60 \times 10^{-19} \times 64.0}}$$

$$= 1.53 \times 10^{-10} \text{ m} = 1.53 \text{ Å}.$$

(1b) Consider the interference between the atomic rows on the surface as shown in Fig. 2 - 6.

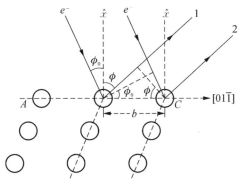

Fig. 2 - 6.

Scattered electron waves from different rows of atoms must interfere constructively if diffracted electrons are to be observable. Hence, the path difference Δl between paths 1 and 2 must be integer multiples of the wavelength so that

$$\Delta l = b(\sin \phi - \sin \phi_0) = n\lambda.$$

Given $\phi_0 = 15.0°$, $\lambda = 1.53$ Å, and $b = \dfrac{a}{\sqrt{2}} = \dfrac{3.92}{\sqrt{2}} = 2.77$ Å, the following two solutions are possible.

(i) When $n = 0$, $\phi = \phi_0 = 15.0°$.

(ii) When $n = 1$, $\Delta l = 2.77(\sin \phi - \sin 15°) = 1 \times 1.53$. Therefore

$$\sin \phi = \frac{1.53 + 0.72}{2.77} = 0.812 \qquad \text{or} \qquad \phi = 54.3°.$$

For $n \geqslant 2$, no solution exists as $\Delta l = 2.77(\sin \phi - \sin 15°) = n \times 1.53$ implies

$$\sin \phi = \sin 15° + \frac{\Delta l}{2.77} = \frac{0.72 + n \times 1.53}{2.77}$$

$$\geqslant \frac{0.72 + 2 \times 1.53}{2.77} = 1.36 > 1.$$

(2) We have $I = I_0 \exp\langle -(\boldsymbol{u} \cdot \Delta \boldsymbol{K})^2 \rangle$.

For elastically scattered electrons, the wavelengths before and after scattering must be equal so that $K' = K$. Therefore, for the specularly reflected beam, the triangle in Fig. 2 - 7 must be an isosceles such that

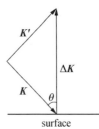

$$\Delta \boldsymbol{K} = \boldsymbol{K}' - \boldsymbol{K} = 2K \cos \theta \, \boldsymbol{x}$$

Fig. 2 - 7.

where $\boldsymbol{x}$ is the unit vector in the direction of the surface normal. Taking the x-component of $\boldsymbol{u}$, we then obtain

$$I = I_0 e^{-\langle u_x^2(t) \cdot 4K^2 \cos^2 \theta \rangle} = I_0 e^{-4K^2 \cos^2 \theta \langle u_x^2(t) \rangle}. \tag{2}$$

The vibration in the direction of the surface normal of a surface atom is assumed to be simple harmonic. Taking $u_x(t) = A \cos \omega t$ and computing the average of the square, we find

$$\langle u_x^2(t) \rangle = \frac{1}{\tau} \int_0^\tau u^2 \, \mathrm{d}t = \frac{1}{\tau} \int_0^\tau A^2 \cos^2 \omega t \, \mathrm{d}t = \frac{A^2}{\tau} \cdot \frac{\tau}{2} = \frac{A^2}{2}.$$

This gives

$$A^2 = 2\langle u_x^2(t) \rangle.$$

The total energy E is thus given by

$$E = \frac{1}{2} C A^2 = \frac{1}{2} C \cdot 2 \langle u_x^2(t) \rangle$$
$$= C \langle u_x^2(t) \rangle = m' \omega^2 \langle u_x^2(t) \rangle.$$

Therefore, one obtains

$$\langle u_x^2(t) \rangle = \frac{E}{m' \omega^2},$$
$$E = m' \omega^2 \langle u_x^2(t) \rangle = k_B T,$$

where m' is the mass of an atom. From either of the above two equations, one then has the following equality

$$\langle u_x^2(t) \rangle = \frac{k_B T}{m' \omega^2} = \frac{k_B T}{m' 4\pi^2 f^2}. \tag{3}$$

From Eqs. (3) and (2), one obtains

$$I = I_0 e^{-4K^2 \cos^2 \theta \frac{k_B T}{m' 4\pi^2 f^2}},$$

where $K = \frac{2\pi p}{h} = \frac{2\pi}{\lambda}$. Accordingly,

$$I = I_0 e^{-\left(\frac{4k_B \cos^2 \theta}{m' f^2 \lambda^2}\right) T} = I_0 e^{-MT} \tag{4}$$

and

$$\ln \frac{I}{I_0} = -M' T,$$

where the slope M' is given by

$$M' = \frac{4 k_B \cos^2 \theta}{m' f^2 \lambda^2}. \tag{5}$$

(a) The slope can be estimated from the plot in Fig. 2 – 5 and leads to the result

$$M' = 2.3 \times 10^{-3} \text{ K}^{-1}.$$

Using the following data in Eq. (5),

$$k_B = 1.38 \times 10^{-23} \text{ J/K},$$
$$\lambda = 1.53 \times 10^{-10} \text{ m},$$

$$m' = 195.\, 1 \times 10^{-3}/(6.\, 02 \times 10^{23}) = 3.\, 24 \times 10^{-25} \text{ kg/atom},$$

one finds

$$2.\, 3 \times 10^{-3} = \frac{4 \times 1.\, 38 \times 10^{-23} \times \cos^2 15°}{(3.\, 24 \times 10^{-25}) \times f^2 \times (1.\, 53 \times 10^{-10})^2}.$$

Thus $f^2 = 3.\, 0 \times 10^{24}$ and the frequency is given by

$$f = 1.\, 7 \times 10^{12} \text{ Hz.}$$

(b) From $\langle u_x^2(t) \rangle = \dfrac{k_B T}{m' 4\pi^2 f^2}$, $T = 300$ K, one finally obtains

$$\langle u_x^2(t) \rangle = \frac{1.\, 38 \times 10^{-23} \times 300}{(3.\, 24 \times 10^{-25}) \times 4\pi^2 \times 3.\, 0 \times 10^{24}} = 1.\, 1 \times 10^{-22} \text{ m}^2$$

and

$$\sqrt{\langle u_x^2(t) \rangle} = 1.\, 0 \times 10^{-11} \text{ m} = 0.\, 10 \text{Å.}$$

Experimental Competition

April 26, 2001   Time Available: 5 hours

Basic Characteristics of Solar Cells

I. Background description

The purpose of this experiment is to explore the basic characteristics of solar cells. Solar cells can absorb electromagnetic waves and convert the absorbed photon energy into electrical energy. A solar cell mainly consists of a diode, whose forward dark current-voltage relationship (i. e. $I - V$ curve under no light illumination) can be expressed as

$$I = I_0(e^{\beta V} - 1),$$

where I_0 and β are constants.

The diode is made up of a semiconductor with a band gap of $E_C - E_V$ (see Fig. 2 - 8). When the energy of the incident photon is larger than the band gap, the photon can be absorbed by the semiconductor to create an electron-hole pair. The electrons and holes are then driven by the internal electric field in the diode to produce a photocurrent (light-generated current). There are several important parameters other than light-generated current involved in a solar cell.

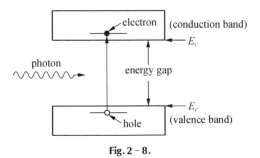

Fig. 2 - 8.

Brief explanations of the terminologies and basic principles are listed below:

(1) Short-circuit current (I_{sc}) is the output current of the solar cell when the external circuit is shorted, i. e. zero load resistance.

(2) Open-circuit voltage (V_{oc}) is the output voltage of the solar cell when the external circuit is open, i. e. infinite load resistance. V_{oc} is also referred to as photovoltaic voltage.

(3) P_m is the maximum output power of the solar cell, i. e. the maximum value of $I \times V$.

(4) The filling factor (FF) is defined to be $\dfrac{P_m}{I_{sc}V_{oc}}$, which represents an important parameter used to evaluate the quality of the solar cell.

(5) Because the photocurrent is produced by photon absorption by the semiconductor, the spectral response of the photocurrent can be used to determine the semiconductor band gap. From the band gap value, one can infer the particular semiconductor material used.

(6) Any incident photon with photon energy larger than the semiconductor band gap can contribute to the photocurrent (I_{ph}) of the solar cell, thus

$$I_{ph} \propto \int_{\lambda_c}^{\lambda_0} N(\lambda) d\lambda,$$

where $N(\lambda)$ is the number of electrons per unit wavelength produced by photons with wavelength λ, λ_c is the cut-off wavelength of the optical filter (see Fig. 2 – 9 ～ Fig. 2 – 11), and λ_0 is the longest wavelength capable of producing a photocurrent. Here $N(\lambda)$ is approximately constant in the visible spectral range, and each optical filter provided in this experiment cuts off all light with wavelengths shorter than a certain cut-off wavelength λ_c. Therefore, the spectral response of I_{ph} with an optical filter can be simplified to be

$$I_{ph} \propto (\lambda_0 - \lambda_c).$$

(7) The photon energy E is related to the photon wavelength as $E = \dfrac{1240}{\lambda}$; the unit for λ is nm (10^{-9} m) and the unit for E is eV (electron-volt) in this equation.

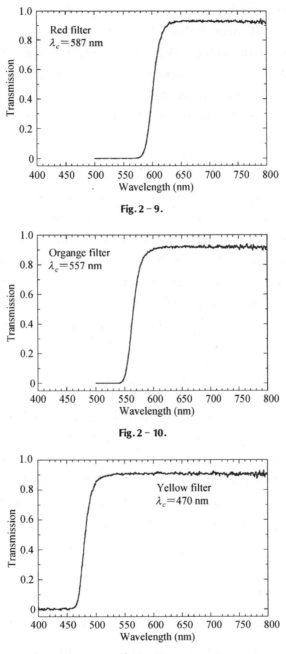

Fig. 2 - 9.

Fig. 2 - 10.

Fig. 2 - 11.

II. Equipments and materials

(1) A solar cell in a black box with pre-installed electrical connections.

(2) Two digital electrical multimeters.

(3) A set of dry cell batteries (1. 5 V × 2).

(4) One precision variable resistor (0 − 5 kΩ).

(**Caution**: Do not connect the central connecting lead (the red wire) of the variable resistor directly to the battery, it may damage the resisitor.)

(5) A white light source with power supply.

(6) Two polarizers. (Note: These polarizers are less effective for light with wavelengths shorter than that of yellow light.)

(7) Red, orange, and yellow optical filters, one of each (their spectral specifications are shown in Fig. 2 − 9∼Fig. 2 − 11).

(8) An optical mount for the optical filters or polarizers. (Note: Optical filters and polarizers can be mounted together on the same optical mount.)

(9) One optical bench.

(10) Wire joining devices: 6 small springs.

(11) One 45 cm ruler.

(12) Regular graph paper (10 sheets), semi-log graph paper (5 sheets).

(13) Two light-shielding boards.

Note: To avoid deterioration due to heat, polarizers and filters should be set at a distance as far away from the light source as possible.

III. Experimental steps

(1) Measure the dark $I - V$ characteristic of the **forward biased** solar cell.

(a) Draw a diagram of the electrical circuit you used.

(b) Plot the $I - V$ curve and determine the values of β and I_0 using the $I - V$ data you obtained.

(2) Measure the characteristics of the solar cell, without electrical bias under white light illumination. (Note: The distance between the light source and the solar cell box should be kept at 30 cm as shown in

Fig. 2 – 12.)

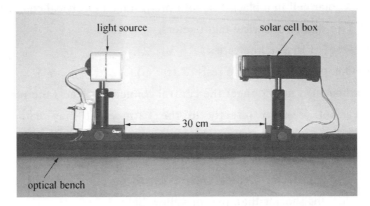

Fig. 2 – 12.

(a) Draw the circuit diagram you used.

(b) Measure the short-circuit current, I_{sc}.

(c) Measure the open-circuit voltage, V_{oc}.

(d) Measure the I vs. V relationship of the solar cell with varying load resistance and plot the $I - V$ curve.

(e) Determine the maximum output power of the solar cell.

(f) Determine the load resistance for the maximum output power.

(g) Calculate the filling factor, $FF = \dfrac{P_m}{I_{sc}V_{oc}}$.

(3) Assume that the solar cell can be modeled as a device consisting of an ideal current source (light-generated current source), an ideal diode, a shunt resistance R_{sh}, and a series resistance R_s.

(a) Draw an equivalent circuit diagram for the solar cell under light illumination.

(b) Derive the $I - V$ relationship for the equivalent circuit. Express the result in terms of R_{sh}, R_s, I_{ph} (light-generated current), and I_d (the current passing through the diode).

(c) Assuming that R_{sh} = infinity and $R_s = 0$ and can be neglected, find the $I - V$ relationship and prove that it can be written in the form as given below:

$$V_{oc} = \beta^{-1}\ln\left(\frac{I_{sc}}{I_0} + 1\right),$$

where V_{oc} is the open-circuit voltage, I_{sc} is the short-circuit current, and I_0, β are constants.

(4) Find effects of irradiance.

(a) Measure and plot the I_{sc} vs. relative light intensity curve, and determine the approximate functional relationship between I_{sc} and the relative light intensity.

(b) Measure and plot the V_{oc} vs. relative light intensity curve, and determine the approximate functional relationship between V_{oc} and the relative light intensity.

(5) Find the wavelength response of the solar cell.

(a) Measure and plot the I_{sc} vs. different cut-off wavelengths using the three optical filters.

(b) Estimate the longest wavelength for which the solar cell can function properly.

(c) Infer which semiconductor material the solar cell is made of. (Hint: The band gaps for commonly used semiconductors are InAs: 0.36 eV, Ge: 0.67 eV, Si: 1.1 eV, amorphous Si(a-Si: H): 1.7 eV, GaN: 3.5 eV.)

Solution

(1) Measure the dark $I - V$ characteristic of the forward biased solar cell.

(a) Circuit diagram

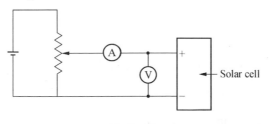

Fig. 2 – 13.

(b) Determine the values β and I_0 in the equation of dark current – voltage characteristic.

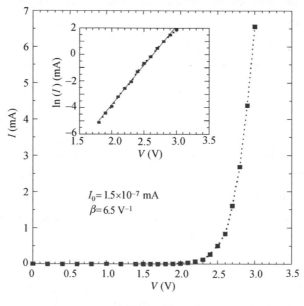

Fig. 2 – 14.

(2) Measure the characteristics of the solar cell, without electrical bias under white light illumination.

(a) Draw the circuit used.

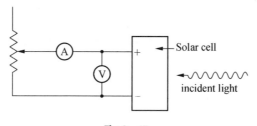

Fig. 2 – 15.

(b) Measure the short-circuit current, I_{sc}.

$$I_{sc} = 2.0 \text{ mA} \sim 4.5 \text{ mA}.$$

(c) Measure the open-circuit voltage, V_{oc}.

$$V_{oc} = 2.7 \text{ V} \sim 3.2 \text{ V}.$$

(d) Measure the I vs. V relationship of the solar cell with varying load resistance and plot the $I - V$ curve.

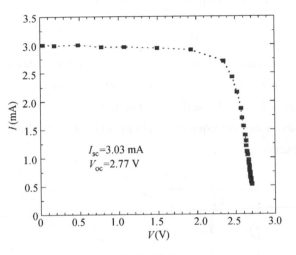

Fig. 2 – 16.

(e) Determine the maximum output power of the solar cell.

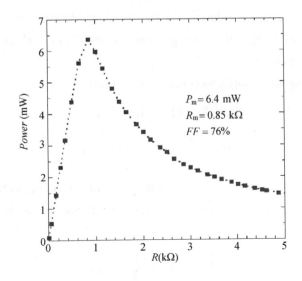

Fig. 2 – 17.

(f) Determine the load resistance for the maximum output power.

$$0.6 \text{ k}\Omega < R < 1.3 \text{ k}\Omega.$$

(g) Calculate the filling factor, $FF \equiv \dfrac{P_m}{I_{sc} V_{oc}}$.

$$62\% < FF < 81\%.$$

(3) Assume that the solar cell can be modeled as a device consisting of an ideal current source (light-generated current source), an ideal diode, a shunt resistance R_{sh}, and a series resistance R_s.

(a) Draw a correct equivalent circuit diagram for the solar cell under light illumination.

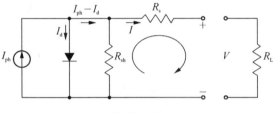

Fig. 2 - 18.

(b) Derive I - V relationship for the equivalent circuit. Express the result in terms of R_{sh}, R_s, light generated current I_{ph}, and current passing through the diode I_d.

Applying Kirchhoff's loop rule, one obtains

$$IR_s + V - (I_{ph} - I_d - I)R_{sh} = 0,$$

$$I\left(1 + \frac{R_s}{R_{sh}}\right) = I_{ph} - \frac{V}{R_{sh}} - I_d.$$

(c) Assuming that $R_{sh} = \infty$ and $R_s = 0$ and can be neglected, find the I-V relationship and prove that it can be written in the form as given below:

$$V_{oc} = \beta^{-1} \ln\left(\frac{I_{sc}}{I_0} + 1\right),$$

where V_{oc} is the open-circuit voltage, I_{sc} is the short-circuit current, and I_0,

β are constants.

Proof. With $R_s = 0$, $R_{sh} = $ infinity, the equivalent circuit becomes

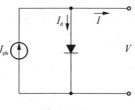

Fig. 2 - 19.

Hence $I = I_{ph} - I_d = I_{ph} - I_0(e^{\beta V} - 1)$.

For close circuit $V = 0$, $I_{ph} = I_{sc}$.

For open circuit $I = 0$, $I_{sc} - I_0(e^{\beta V_{oc}} - 1) = 0$

$$\therefore V_{oc} = \frac{1}{\beta}\ln\left(\frac{I_{sc}}{I_0} + 1\right).$$

(4) Find effects of irradiance.

(a) Measure and plot the I_{sc} vs. relative light intensity curve, and determine the approximate functional relationship between I_{sc} and the relative light intensity.

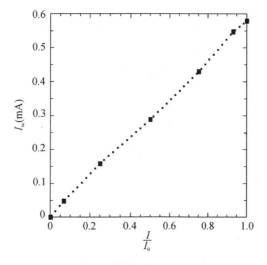

Fig. 2 - 20.

I_0 is the light intensity when the axes of the two polarizers are parallel.

Because polarizers are less effective for light with wavelength shorter than that of yellow light as mentioned in the description of "Equipments and Materials", in order to obtain correct relative scale of irradiance, the yellow filter should be used with polarizers.

I_{sc} is approximately proportional to light intensity, $I_{sc} = A \dfrac{I}{I_0}$.

(b) Measure and plot the V_{oc} vs. relative light intensity curve, and determine the approximate functional relationship between V_{oc} and the relative light intensity.

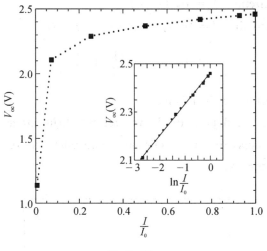

Fig.2 - 21.

$$V_{oc} = A \ln \frac{I}{I_0} + B \quad \text{or} \quad V_{oc} = A' \ln \frac{I}{I_0} + B'.$$

(5) Find the wavelength response of solar cell under different irradiance.

(a) Measure and plot the I_{sc} vs. different cut-off wavelengths using the three optical filters.

(b) Estimate the longest wavelength for which the solar cell can function properly.

$\lambda_0 = 730 \sim 770$ nm.

(c) Infer which semiconductor material the solar cell is made of. (Hint: The band gaps for commonly used semiconductors are InAs:

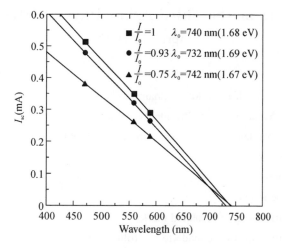

Fig. 2 - 22.

0. 36 eV, Ge: 0. 67 eV, Si: 1. 1 eV, amorphous Si (a-Si: H): 1. 7 eV, GaN: 3. 5 eV).

The solar cell is made of a-Si: H.

Minutes of the Third Asian Physics Olympiad

April 12, 2002

Present: A/P Cheah Horn Mun [Chairman]

Dr Yohanes Surya [President, APhO]

Dr Waldemar Gorzkowski [President, IPhO]

Leader

Observers

Item	Description
1	Corrections to tabulated marks
1. 1	Corrections were made to the following:

- No. 64: Zhang Feng-the correct marks should be 35 marks, hence a gold medal instead of silvers is awarded
- No. 82: Weerapong Phadungsukanan-was awarded 25 marks, therefore should be awarded with a bronze medal instead of an Honorable Mention

1. 2	All the leaders and observers accepted the corrections.
2	Prizes
2. 1	The Best Overall Prize was awarded to Gu Chun Chui (People's Republic of China).
2. 2	The Best Experiment Prize was awarded to Fan Xiangjun (People's Republic of China).
2. 3	The Best Theory Prize went to Agustinus Peter Sahanggamu (Indonesia).
2. 4	The Most Creative Solution Prize went to Thaned Pruttivarsin (Thailand).
2. 5	The Best Newcomer Prize was won by Giga Chkareuli (Georgia).
2. 6	The total number of participants was 105. All results were accepted by the leaders and observers through voting.
3	Guest-of-honour for closing ceremony
3. 1	Chairman informed that the Minister of Education, Dr Ng Eng Hen, would be the Guest-of-Honor for the closing ceremony. In addition, the Thai Deputy Minister for Education would also be attending the

ceremony.

4	Australia as member of Apho
4. 1	IPhO President proposed for the statute to include and recognize the acceptance of Australia into the Asian Physics Olympiad.
4. 2	The new statute would read:

By the term "countries of the Asian region" one should understand countries whose capitals are localized in the region traditionally recognized as Asia and Australia.

4. 3	All changes to the statute are to be disseminated to all countries three months in advance.
4. 4	The meeting accepts the change in the statute with a total of 26 votes.
5	Organisers of APhO
5. 1	APhO President expressed his appreciation to the organizing committee for organizing APhO 2002.
5. 2	Future hosts of APhO, based on preliminary agreement, are as follows:

- 2003 – Thailand
- 2004 – Vietnam
- 2005 – Indonesia
- 2006 – Georgia
- 2007 – People's Republic of China (tentative)
- 2008 – Australia (tentative)
- 2009 – Uzbekistan (tentative)
- 2010 – Malaysia (tentative)
- 2011 – Israel (tentative)

6	Equipment for experiment exam
6. 1	Chairman informed that the equipment used for the experiment exam would be distributed to the teams as souvenirs. The equipment would be delivered to the leaders' rooms on the same night.

As there is no any other matter, the meeting ended at 6. 00 p. m.

Recorded by: Miss Nenny Aryanti Noorman.

Theoretical Competition

May 8, 2002   Time available: 5 hours

Problem 1
Vibrations of a linear crystal lattice

A very large number N of movable identical point particles $(N \gg 1)$, each with mass m, are set in a straight chain with $N + 1$ identical massless springs, each with stiffness S, linking them to each other and the ends attached to two additional immovable particles. See Fig. 3 – 1. This chain will serve as a model of the vibration modes of a one – dimensional crystal. When the chain is set in motion, the longitudinal vibrations of the chain can be looked upon as a superposition of simple oscillations (called modes) each with its own characteristic mode frequency.

$$\begin{array}{cccccccc} & S & m & m & & S & & m & S & m & S \\ 0 & a & 2a & 3a & & na & & (N-1)a & Na & (N+1)a = L \end{array}$$

Fig. 3 – 1.

(a) Write down the equation of motion of the nth particle.

(b) To attempt to solve the equation of motion of part (a) use the trial solution

$$x_n(\omega) = A \sin nka \cos(\omega t + \alpha),$$

where $x_n(\omega)$ is the displacement of the nth particle from equilibrium, ω the angular frequency of the vibration mode and A, k and α are constants; k and ω are the wave numbers and mode frequencies respectively. For each k, there will be a corresponding frequency ω. Find the dependence of ω on k, the allowed values of k, and the maximum value of ω. The chain's vibration is thus a superposition of all these vibration modes. Useful formulas:

$$\left(\frac{\mathrm{d}}{\mathrm{d}x}\right) \cos \alpha x = -\alpha \sin \alpha x, \quad \left(\frac{\mathrm{d}}{\mathrm{d}x}\right) \sin \alpha x = \alpha \cos \alpha x, \quad \alpha = \text{constant};$$

$$\sin(A+B) = \sin A \cos B + \cos A \sin B;$$

$$\cos(A+B) = \cos A \cos B - \sin A \sin B.$$

According to Planck the energy of a photon with a frequency of ω is $\hbar\omega$, where $\hbar$ is the Planck constant divided by 2π. Einstein made a leap from this by assuming that a given crystal vibration mode with frequency ω also has this energy. Note that a vibration mode is not a particle, but a simple oscillation configuration of the entire chain. This vibration mode is analogous to the photon and is called a phonon. We will follow up the consequences of this idea in the rest of the problem. Suppose a crystal is made up of a very large ($\sim 10^{23}$) number of particles in a straight chain.

(c) For a given allowed ω (or k) there may be no phonons; or there may be one; or two; or any number of phonons. Hence it makes sense to try to calculate the average energy $\langle E(\omega) \rangle$ of a particular mode with a frequency ω. So $\langle E(\omega) \rangle$ is the average energy of a phonon with frequency ω. Let $P_p(\omega)$ represent the probability that there are p phonons with this frequency ω. Then the required average is

$$\langle E(\omega) \rangle = \frac{\displaystyle\sum_{p=0}^{\infty} p\hbar\omega \, P_p(\omega)}{\displaystyle\sum_{p=0}^{\infty} P_p(\omega)}.$$

Although the phonons are discrete, the fact that there are so many of them (and the P_p becomes tiny for large p) allows us to extend the sum to $p = \infty$, with negligible error. Now the probability P_p is given by Boltzmann's formula

$$P_p(\omega) \propto \exp\left(\frac{-p\hbar\omega}{k_B T}\right),$$

where k_B is Boltzmann's constant and T is the absolute temperature of the crystal, assumed constant. The constant of proportionality does not depend on p. Calculate the average energy for phonons of frequency ω. Possibly useful formula:

$$\frac{\mathrm{d}}{\mathrm{d}x} e^{f(x)} = \frac{\mathrm{d}f}{\mathrm{d}x} e^{f(x)}.$$

(d) We would like next to compute the total energy E_T of the crystal. In part (c) we found the average energy $\langle E(\omega) \rangle$ for the vibration mode ω. To find E_T we must multiply $\langle E(\omega) \rangle$ by the number of modes of the crystal with frequency ω and then sum up all these for the entire range from $\omega = 0$ to ω_{max}. Take an interval Δk in the range of wave numbers. For very large N and for Δk much larger than the spacing between successive (allowed) k values, how many modes can be found in the interval Δk?

(e) To make use of the results of (a) and (b), approximate Δk by $\dfrac{dk}{d\omega} d\omega$ and replace any sum by an integral over ω. (It is more convenient to use the variable ω in place of k at this point.) State the total number of modes of the crystal in this approximation. Also derive an expression E_T but do not evaluate it. The following integral may be useful:

$$\int_0^1 \frac{dx}{\sqrt{1-x^2}} = \frac{\pi}{2}.$$

(f) The molar heat capacity C_V of a crystal at constant volume is experimentally accessible: $C_V = \dfrac{dE_T}{dT}$ (T = absolute temperature). For the crystal under discussion determine the dependence of C_V on T for very large and very low temperatures (i. e. is it constant, linear or power dependent for an interval of the temperature?). Sketch a qualitative graph of C_V versus T, indicating the trends predicted for very low and very high T.

Solution 1

(a) $m \ddot{x}_n = S(x_{n+1} - x_n) - S(x_n - x_{n-1})$.

(b) Let $x_n = A \sin nka \, \cos(\omega t + \alpha)$, which has a harmonic time dependence.

By analogy with the spring, the acceleration is $\ddot{x}_n = -\omega^2 x_n$.

Substitute into (a):

$$-mA\omega^2 \sin nka = AS[\sin(n+1)ka - 2 \sin nka + \sin(n-1)ka]$$

$$= -4SA \sin nka \, \sin^2 \frac{1}{2}ka.$$

$$\text{Hence } \omega^2 = \frac{4S}{m}\sin^2\frac{1}{2}ka.$$

To determine the allowed values of k, use the boundary condition

$$\sin(N+1)ka = \sin kL = 0.$$

The allowed wave numbers are given by

$$kL = \pi,\ 2\pi,\ 3\pi,\ \ldots,\ N\pi \ (N \text{ in all}),$$

and their corresponding frequencies can be computed from $\omega = \omega_0 \sin\frac{1}{2}ka$,

in which $\omega_{\max} = \omega_0 = 2\left(\dfrac{S}{m}\right)^{\frac{1}{2}}$ is the maximum allowed frequency.

(c) $\langle E(\omega)\rangle = \dfrac{\displaystyle\sum_{p=0}^{\infty} p\hbar\omega P_{\mathrm p}(\omega)}{\displaystyle\sum_{p=0}^{\infty} P_{\mathrm p}(\omega)}.$

First method:

$$\frac{\displaystyle\sum_{n=0}^{\infty} p\hbar\omega\, e^{-\frac{n\hbar\omega}{k_{\mathrm a}T}}}{\displaystyle\sum_{n=0}^{\infty} e^{-\frac{p\hbar\omega}{k_{\mathrm a}T}}} = k_B T^2 \frac{\partial}{\partial T}\ln\sum_{n=0}^{\infty} e^{-\frac{p\hbar\omega}{k_{\mathrm a}T}}.$$

The sum is a geometric series and is $(1 - e^{-\frac{\hbar\omega}{k_{\mathrm a}T}})^{-1}$.

We find $\langle E(\omega)\rangle = \dfrac{\hbar\omega}{e^{\frac{\hbar\omega}{k_{\mathrm a}T}} - 1}.$

Alternatively: denominator is a geometric series $= (1 - e^{-\frac{\hbar\omega}{k_{\mathrm a}T}})^{-1}$.

Numerator is $k_B T^2 \dfrac{\mathrm d}{\mathrm dT}$ (denominator) $= e^{-\frac{\hbar\omega}{k_{\mathrm a}T}}(1 - e^{-\frac{\hbar\omega}{k_{\mathrm a}T}})^{-2}$ and result follows.

A non-calculus method:

Let $D = 1 + e^{-x} + e^{-2x} + e^{-3x} + \ldots$, where $x = \dfrac{\hbar\omega}{k_B T}$. This is a geometric series and equals $D = \dfrac{1}{1 - e^{-x}}$. Let $N = e^{-x} + 2e^{-2x} + 3e^{-3x} + \ldots$. The result we want is $\dfrac{N}{D}$. Observe

$$D-1 = e^{-x} + e^{-2x} + e^{-3x} + e^{-4x} + e^{-5x} + \dots$$
$$(D-1)e^{-x} = e^{-2x} + e^{-3x} + e^{-4x} + e^{-5x} + \dots$$
$$(D-1)e^{-2x} = e^{-3x} + e^{-4x} + e^{-5x} + \dots.$$

Hence $N = (D-1)D$ or $\dfrac{N}{D} = D-1 = \dfrac{e^{-x}}{1-e^{-x}} = \dfrac{1}{e^{x}-1}$.

(d) From part (b), the allowed k values are $\dfrac{\pi}{L}, \dfrac{2\pi}{L}, \dots, \dfrac{N\pi}{L}$.

Hence the spacing between allowed k values is $\dfrac{\pi}{L}$, so there are $\dfrac{L}{\pi}\Delta k$ allowed modes in the wave-number interval Δk $\left(\text{assuming } \Delta k \gg \dfrac{\pi}{L}\right)$.

(e) Since the allowed k are $\dfrac{\pi}{L}, \dots, \dfrac{N\pi}{L}$, there are N modes.

Follow the problem:

$$\frac{d\omega}{dk} = \frac{1}{2}a\omega_0 \cos \frac{1}{2}ka \text{ from parts (a) and (b)}$$

$$= \frac{1}{2}a\sqrt{\omega_{max}^2 - \omega^2}, \ \omega_{max} = \omega_0.$$

This second form is more convenient for integration.
The number of modes dn in the interval $d\omega$ is

$$dn = \frac{L}{\pi}\Delta k = \frac{L}{\pi}\frac{dk}{d\omega}d\omega$$

$$= \frac{L}{\pi}\left(\frac{1}{2}a\omega_0 \cos \frac{1}{2}ka\right)^{-1}d\omega$$

$$= \frac{L}{\pi}\frac{2}{a}\frac{1}{\sqrt{\omega_{max}^2 - \omega^2}}d\omega$$

$$= \frac{2(N+1)}{\pi}\frac{1}{\sqrt{\omega_{max}^2 - \omega^2}}d\omega.$$

Total number of modes $= \displaystyle\int dn = \int_0^{\omega_{max}} \frac{2(N+1)}{\pi}\frac{d\omega}{\sqrt{\omega_{max}^2 - \omega^2}} = N+1 \approx$

N for large N.

Total crystal energy from (c) and dn of part (e) is given by

$$E_T = \frac{2N}{\pi}\int_0^{\omega_{max}} \frac{\hbar\omega}{e^{\frac{\hbar\omega}{k_B T}} - 1}\frac{d\omega}{\sqrt{\omega_{max}^2 - \omega^2}}.$$

(f) Observe first from the last formula that E_T increases monotonically with temperature since $(e^{\frac{\hbar\omega}{kT}} - 1)^{-1}$ is increasing with T.

When $T \to 0$, the term -1 in the last result may be neglected in the denominator so

$$E_T \approx_{T \to 0} \frac{2N}{\pi} \int \hbar\omega e^{-\frac{\hbar\omega}{k_B T}} \frac{1}{\sqrt{\omega_{max}^2 - \omega^2}} d\omega$$

$$= \frac{2N}{\hbar \pi \omega_{max}} (k_B T)^2 \int_0^\infty \frac{x e^{-x}}{\sqrt{1 - \left(\frac{k_B T x}{\hbar\omega_{max}}\right)^2}} dx,$$

which is quadratic in T (denominator in integral is effectively unity) hence C_V is linear in T near absolute zero.

Alternatively, if the summation is retained, we have

$$E_T = \frac{2N}{\pi} \sum_\omega \frac{\hbar\omega}{e^{\frac{\hbar\omega}{k_B T}} - 1} \frac{\Delta\omega}{\sqrt{\omega_{max}^2 - \omega^2}}.$$

When $T \to 0$,

$$E_T \approx \frac{2N}{\pi} \sum_\omega \hbar\omega e^{-\frac{\hbar\omega}{k_B T}} \frac{\Delta\omega}{\sqrt{\omega_{max}^2 - \omega^2}}$$

$$= \frac{2N}{\pi} \frac{(k_B T)^2}{\hbar\omega} \sum_y e^{-y} y \Delta y.$$

When $T \to \infty$, use $e^x \approx 1 + x$ in the denominator,

$$E_T \approx_{T \to \infty} \frac{2N}{\pi} \int_0^{\omega_{max}} \frac{\hbar\omega}{\frac{\hbar\omega}{k_B T}} \frac{1}{\sqrt{\omega_{max}^2 - \omega^2}} d\omega = \frac{2N}{\pi} k_B T \frac{\pi}{2},$$

which is linear; hence $C_V \to Nk_B = R$, the universal gas constant. This is the Dulong – Petit rule.

Alternatively, if the summation is retained, write denominator as $e^{\frac{\hbar\omega}{k_B T}} - 1 \approx \frac{\hbar\omega}{k_B T}$ and when $T \to \infty$,

$$E_T \approx \frac{2N}{\pi} k_B T \sum_\omega \frac{\Delta\omega}{\sqrt{\omega_{max}^2 - \omega^2}},$$

which is linear in T, so C_V is constant.

Sketch of C_V versus T:

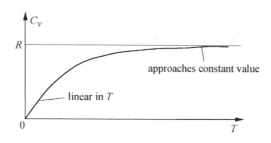

Fig. 3 – 2.

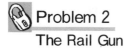

Problem 2
The Rail Gun

A young man at P and a young lady at Q were deeply in love. These two places are separated by a strait of width $w = 1000$ m. After learning about the theory of rail gun in class, the young man could not wait to construct such a device to launch himself across the strait. He constructed a ramp of adjustable elevation of angle θ on which he laid two metal rails (the length of each rail is $D = 35.0$ m) in parallel, separated by $L = 2.00$ m. He managed to connect a 2424 V DC power supply to the ends of the rails. A conducting bar can slide freely on the metal rails such that he could hang on to it safely as it slides.

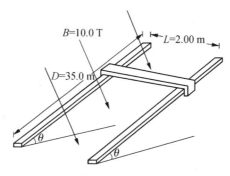

Fig. 3 – 3.

A skilled engineer, moved by all these efforts, designed a system that can produce a $B = 10.0$ T magnetic field that can be directed perpendicular to the plane of the rails. The mass of the young man is 70 kg. The mass of the conducting bar is 10 kg and its resistance is $R = 1.0\ \Omega$.

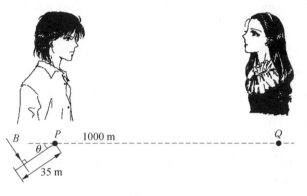

Fig. 3 – 4.

Just after he had completed the construction and checked that it worked perfectly, he received a call from the young lady, sobbing and telling him that her father was going to marry her off to a rich man unless he can arrive at Q within 11 seconds after the call, and having said that she hang up.

The young man immediately got into action and launched himself across the strait to Q.

Show, using the steps listed below, whether it is possible for him to make it in time, and if so, what is the range of θ he must set the ramp?

(a) Derive an expression for the acceleration of the young man parallel to the rail.

(b) Obtain an expression in terms of θ for the time spent

(i) on the rails, t_s and

(ii) in flight, t_f.

(c) Plot a graph of the total time $T = t_s + t_f$ against the angle of inclination θ.

(d) By considering the relevant parameters of this device, obtain the range of angles that he should set. Plot another graph if necessary.

Make the following assumptions:

(1) The time between the end of the call and all preparations (such as

setting θ to the appropriate angle) for the launch is negligible. This is to say, the launch is considered to start at time $t = 0$ when the bar (with the young man hanging to it) is starting to move.

(2) The young man may start his motion from any point along the metal rails.

(3) The higher end of the ramp and Q is at the same level, and the distance between them is $w = 1000$ m.

(4) There is no question about safety such as when landing, electric shocks, etc.

(5) The resistance of the metal rails, the internal resistance of the power supply, the friction between the conducting bar and the rails and the air resistance are all negligible.

(6) Take acceleration due to gravity as $g = 10$ m/s^2.

Some Mathematical Notes:

(1) $\int e^{-ax} \, dx = -\dfrac{e^{-ax}}{a}$.

(2) The solution to $\dfrac{dx}{dt} = a + bx$ is given by

$$x(t) = \frac{a}{b}(e^{bt} - 1) + x(0)e^{bt}.$$

Solution

Proper Solution (taking induced emf into consideration):

(a) Let I be the current supplied by the battery in the absence of back emf.

Let i be the induced current by back emf ε_b.

Since $\varepsilon_b = \dfrac{d\phi}{dt} = \dfrac{d(BLx)}{dt} = BLv, \quad \therefore \quad i = \dfrac{Blv}{R}$.

Net current, $I_N = I - i = I - \dfrac{BLv}{R}$.

Forces parallel to rail are:

Force on rod due to current is

$$F_c = BLI_N = BL\left(I - \frac{BLv}{R}\right)$$

$$= BLI - \frac{B^2L^2v}{R}.$$

Net force on rod and young man combined is $F_N = F_c - mg\sin\theta$. (1)

Newton's law: $\qquad\qquad F_N = ma = \dfrac{m\,dv}{dt}.$ (2)

Equating (1) and (2), substituting for F_c and dividing by m, we obtain the acceleration

$$\frac{dv}{dt} = \alpha - \frac{v}{\tau}, \text{ where } \alpha = \frac{BIL}{m} - g\sin\theta \text{ and } \tau = \frac{mR}{B^2L^2}.$$

(b) (i) Since initial velocity of rod $= 0$, and let velocity of rod at time t be $v(t)$, we have

$$v(t) = v_\infty\left(1 - e^{-\frac{t}{\tau}}\right), \tag{3}$$

where $v_\infty(\theta) = \alpha\tau = \dfrac{IR}{BL}\left(1 - \dfrac{mg}{BLI}\sin\theta\right)$.

Let t_s be the total time he spent moving along the rail, and v_s be his velocity when he leaves the rail, i.e.

$$v_s = v(t_s) = v_\infty\left(1 - e^{-\frac{t_s}{\tau}}\right) \tag{4}$$

$$\therefore t_s = -\tau\ln\left(1 - \frac{v_s}{v_\infty}\right). \tag{5}$$

(b) (ii) Let t_f be the time in flight:

$$t_f = \frac{2v_s\sin\theta}{g}. \tag{6}$$

He must travel a horizontal distance w during t_f.

$$w = (v_s\cos\theta)t_f, \tag{7}$$

$$t_f = \frac{w}{v_s\cos\theta} = \frac{2v_s\sin\theta}{g}.$$

$\qquad\qquad\qquad\qquad\qquad$ (8) (from (6) and (7))

From (8), v_s is fixed by the angle θ and the width of the strait w

$$v_s = \sqrt{\frac{gw}{\sin 2\theta}}, \tag{9}$$

$$\therefore t_s = -\tau \ln\left(1 - \frac{1}{v_\infty}\sqrt{\frac{gw}{\sin 2\theta}}\right),$$

(Substitute (9) in (5))

and
$$t_f = \frac{2\sin\theta}{g}\sqrt{\frac{gw}{\sin 2\theta}} = \sqrt{\frac{2w\tan\theta}{g}}.$$

(Substitute (9) in (8))

(c) Therefore, total time is:

$$T = t_s + t_f = -\tau \ln\left(1 - \frac{1}{v_\infty}\sqrt{\frac{gw}{\sin 2\theta}}\right) + \sqrt{\frac{2w\tan\theta}{g}}.$$

The values of the parameters are: $B = 10.0$ T, $I = 2424$ A, $L = 2.00$ m, $R = 1.0\ \Omega$, $g = 10$ m/s^2, $m = 80$ kg, and $w = 1000$ m.

Then
$$\tau = \frac{mR}{B^2 L^2} = \frac{(80)(1.0)}{(10.0)^2(2.00)^2} = 0.20 \text{ s}.$$

$$v_\infty(\theta) = \frac{2424}{(10.0)(2.00)}\left(1 - \frac{(80)(10)}{(10.0)(2.00)(2424)}\sin\theta\right)$$

$$= 121(1 - 0.0165\sin\theta).$$

So

$$T = t_s + t_f = -0.20\ln\left(1 - \frac{100}{v_\infty}\frac{1}{\sqrt{\sin 2\theta}}\right) + 14.14\sqrt{\tan\theta}.$$

By plotting T as a function of θ, we obtain the following graph:

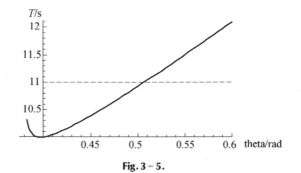

Fig. 3 - 5.

Note that the lower bound for the range of θ to plot may be determined by the condition $\frac{v_s}{v_\infty} < 1$ (or the argument of ln is positive), and since $\frac{mg}{BIL}$ is small (0.0165), $v_\infty \approx \frac{IR}{BL} (= 121\,\text{m/s})$, we have the condition $\sin 2\theta > 0.68$, i. e. $\theta > 0.37$. So one may start plotting from $\theta = 0.38$.

From the graph, for θ within the range $(0.38 \sim 0.505)$ radian the time T is within 11 s.

(d) However, there is another constraint, i. e. the length of rail D. Let D_s be the distance travelled during the time interval t_s

$$D_s = \int_0^{t_s} v(t)\,dt = v_\infty \int_0^{t_s} (1 - e^{-\frac{t}{\tau}})\,dt$$

$$= v_\infty (t + \tau e^{-\beta t})_0^{t_s}$$

$$= v_\infty [t_s - \tau(1 - e^{-\beta t})]$$

$$= v_\infty t_s - v(t_s)\tau,$$

i. e.

$$D_s = -\tau \left[v_\infty (\theta) \ln \left(1 - \frac{1}{v_\infty (\theta)} \sqrt{\frac{gw}{\sin 2\theta}} \right) + \sqrt{\frac{gw}{\sin 2\theta}} \right].$$

The graph below shows D_s as a function of θ.

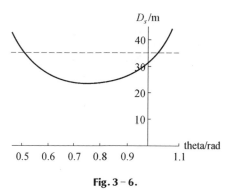

Fig. 3 − 6.

It is necessary that $D_s \leqslant D$, which means θ must range between 0.5 and 1.06 radians.

In order to satisfy both conditions, θ must range between 0.5 and 0.505

radians.

Remarks: Using the formula for t_f, t_s & D, we get

At $\theta = 0.507$, $t_f = 10.540$, $t_s = 0.466$, giving $T = 11.01$ s, $D = 34.3$ m;

At $\theta = 0.506$, $t_f = 10.527$, $t_s = 0.467$, giving $T = 10.99$ s, $D = 34.4$ m;

At $\theta = 0.502$, $t_f = 10.478$, $t_s = 0.472$, giving $T = 10.95$ s, $D = 34.96$ m;

At $\theta = 0.50$, $t_f = 10.453$, $t_s = 0.474$, giving $T = 10.93$ s, $D = 35.2$ m.

So the more precise angle range is between 0.502 to 0.507, but students are not expected to give such answers.

To 2 sig fig $T = 11$ s. Range is 0.50 to 0.51 (in degree: $28.6°$ to $29.2°$ or $29°$).

Alternate Solution (Not taking induced emf into consideration):

If induced emf is not taken into account, there is no induced current, so the net force acting on the combined mass of the young man and rod is

$$F_N = BIL - mg\sin\theta.$$

And we have instead

$$\frac{dv}{dt} = a,$$

where

$$a = \frac{BIL}{m} - g\sin\theta.$$

$$\therefore v(t) = at$$

and

$$\therefore v_s = v(t_s) = at_s$$

$$t_f = \frac{2v_s\sin\theta}{g} = \frac{2at_s\sin\theta}{g}.$$

Therefore,

$$w = (v_s\cos\theta)t_f = \frac{a^2 t_s^2 \sin 2\theta}{g},$$

giving

$$t_s = \frac{1}{a}\sqrt{\frac{gw}{\sin 2\theta}}$$

and
$$t_f = \sqrt{\frac{2w\tan\theta}{g}}.$$

Hence,
$$T = t_s + t_f = \frac{1}{a}\sqrt{\frac{gw}{\sin 2\theta}} + \sqrt{\frac{2w\tan\theta}{g}}$$

$$= \frac{\sqrt{wg}}{a} \frac{\left[1 + 2\left(\dfrac{a}{g}\right)\sin\theta\right]}{\sqrt{\sin 2\theta}},$$

where
$$a = \frac{BIL}{m} - g\sin\theta.$$

The values of the parameters are: $B = 10.0$ T, $I = 2424$ A, $L = 2.00$ m, $R = 1.0\,\Omega$, $g = 10$ m/s^2, $m = 80$ kg, and $w = 1000$ m. Then,

$$T = \frac{100}{a} \frac{[1 + 0.20a\sin\theta]}{\sqrt{\sin 2\theta}},$$

where
$$a = 606 - 10\sin\theta.$$

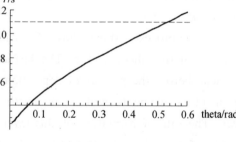

Fig. 3 - 7.

For θ within the range $(0\sim0.52)$ radian the time T is within 11 s. However, there is another constraint, i.e. the length of rail D. Let D_s be the distance travelled during the time interval t_s

$$D_s = \frac{gw}{2a\sin 2\theta} = \frac{5000}{a\sin 2\theta},$$

which is plotted below

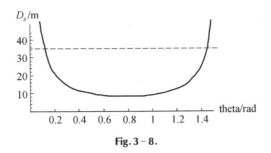

Fig. 3 – 8.

It is necessary that $D_s \leqslant D$, which means θ must range between 0.11 and 1.43 radians.

In order to satisfy both conditions, θ must range between 0.11 and 0.52 radians.

Problem 3
Wafer Fabrication

Wafer fabrication refers to the production of semiconductor chips from silicon. In modern technologies there are more than 20 processes; we are going to concentrate on thin films deposition.

In wafer fabrication process, thin films of various materials are deposited on the surface of the silicon wafer. The surface of the substrate must be extremely clean before the process of deposition. The presence of traces of oxygen or other elements will result in the formation of a contamination layer. The rate of formation of this layer is determined by the impingement rate of the gas molecules hitting the substrate surface. Assuming the number of molecules per unit volume is n, the impingement rate on a unit area of the substrate from the gas is given by

$$J = \frac{1}{4}n\bar{v},$$

where $\bar{v}$ is the average or mean speed of the gas molecules.

(a) Assuming that the gas molecules obey a Maxwell – Boltzmann distribution,

$$W(v) = 4\pi \left(\frac{M}{2\pi RT}\right)^{\frac{3}{2}} v^2 e^{-\frac{Mv^2}{2RT}},$$

where $W(v) \, dv$ is the fraction of molecules whose speed lie between v and $v + dv$, M is the molar mass of the gas, T is the gas temperature and R is the gas constant, show that the average or mean speed of the gas molecules is given by

$$\bar{v} = \int_0^\infty v W(v) \, dv$$
$$= \sqrt{\frac{8RT}{\pi M}}.$$

(b) Assuming that the gases behave as an ideal gas at low pressure, P, show that the rate of impingement is given by

$$J = \frac{P}{\sqrt{2\pi m k T}},$$

where m is the mass of the molecule and T is the temperature of the gas.

(c) If the residual pressure of oxygen in a vacuum system is 133 Pa, and by modelling the oxygen molecule as a sphere of radius approximately 3.6×10^{-10} m, estimate how long it takes to deposit a molecule-thick layer of oxygen on the wafer at $300°$ Celsius, assuming that all the oxygen molecules which strike the silicon wafer surface are deposited. Assume also that oxygen molecules in the layer are arranged side by side.

(d) In reality, not all molecules of oxygen react with the silicon. This can be modeled by the concept of activation energy where the reacting molecules should have total energy greater than the activation energy before it can react. Physically this activation energy describes the fact that chemical bonds between the silicon atoms have to be broken before a new bond between silicon and oxygen atoms is formed. Assuming an activation energy for the reaction to be 1 eV, estimate again how long it would take to deposit one atomic layer of oxygen at the above temperature. You may assume that the area under the Maxwell distribution in part (a) is unity.

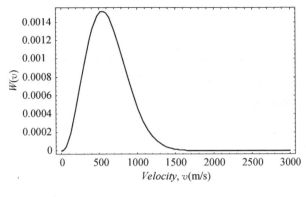

Fig. 3 – 9.

(e) For lithography processes, the clean silicon wafer is coated evenly with a layer of transparent polymer (photo-resist) of refractive index $\mu = 1.40$. To measure the thickness of this photo-resist, the wafer is illuminated with collimated monochromatic beam of light of wavelength $\lambda = 589$ nm. For a certain minimum thickness of photo-resist, d, there is a destructive interference of reflected light, assuming normal incidence on the coating. Derive an expression for relation between d, μ and λ. Calculate d using the given data. In this point you may assume that silicon behaves as a medium with a refractive index greater than 1.40 and you may ignore multiple reflections.

The following data may be helpful:

Molar mass of oxygen is 32 g mol^{-1}.

Boltzmann constant, $k = 1.38 \times 10^{-23}$ JK^{-1}.

Avogadro number, $N_A = 6.02 \times 10^{23}$ mol^{-1}

Useful formula:

$$\int x^3 e^{-kx^2}\, dx = -\frac{1}{2} e^{-kx^2}\left(\frac{1}{k^2} + \frac{x^2}{k}\right).$$

 Solution

(a) Since $W(v) = 4\pi\left[\dfrac{M}{2\pi RT}\right]^{\frac{3}{2}} v^2 e^{-\frac{Mv^2}{2RT}}$,

$$\bar{v} = \int_0^\infty v W(v)\,dv = \int_0^\infty v 4\pi \left(\frac{M}{2\pi RT}\right)^{\frac{3}{2}} v^2 e^{-\frac{Mv^2}{2RT}}\,dv$$

$$= 4\pi \left(\frac{M}{2\pi RT}\right)^{\frac{3}{2}} \int_0^\infty v^3 e^{-\frac{Mv^2}{2RT}}\,dv = 4\pi \left(\frac{M}{2\pi RT}\right)^{\frac{3}{2}} \frac{4R^2 T^2}{2M^2}$$

$$= \sqrt{\frac{8RT}{\pi M}}.$$

(b) Assuming an ideal gas, $pV = NkT$, so that the concentration of the gas molecules, n, is given by

$$n = \frac{N}{V} = \frac{p}{kT},$$

the impingement rate is given by

$$J = \frac{1}{4} n \bar{v} = \frac{1}{4} \frac{p}{kT} \sqrt{\frac{8RT}{\pi M}} = p \sqrt{\frac{8RT}{16 k^2 T^2 \pi M}}$$

$$= p \sqrt{\frac{N_A k}{2kT \pi M}} = p \sqrt{\frac{1}{2kT \pi m}}$$

$$= \frac{p}{\sqrt{2\pi m k T}},$$

where we have note that $R = N_A k$ and $m = \dfrac{M}{N_A}$ (N_A being Avogadro number).

(c) Assuming close packing, there are approximately 4 molecules in an area of $16r^2$ m^2. Thus, the number of molecules in 1 m^2 is given by

$$n_1 = \frac{4}{16(3.6 \times 10^{-10})^2} = 1.9 \times 10^{18} \text{ m}^{-2}.$$

However at $(273 + 300)$K and 133 Pa, the impingement rate for oxygen is

$$J = \frac{p}{\sqrt{2\pi m k T}}$$

$$= \frac{133}{\sqrt{2\pi \left(\dfrac{32 \times 10^{-3}}{6.02 \times 10^{23}}\right)(1.38 \times 10^{-23})573}}$$

$$= 2.6 \times 10^{24} \text{ m}^{-2} \text{s}^{-1}.$$

Therefore, the time needed for the deposition is $\frac{n_1}{J} = 0.7\ \mu s$.

The calculated time is too short compared with the actual processing.

(d) With activation energy of 1 eV and letting the velocity of the oxygen molecule at this energy is v_1, we have

$$\frac{1}{2}mv_1^2 = 1.6 \times 10^{-19}\ \text{J}$$

$$\Rightarrow v_1 = 2453.57\ \text{ms}^{-1}.$$

At a temperature of 573 K, the distribution of the gas molecules is

$$W(v) = 4\pi\left(\frac{32 \times 10^{-3}}{2\pi \times 8.31 \times 573}\right)^{\frac{3}{2}} v^2 \exp\left(-\frac{32 \times 10^{-3}}{2 \times 8.31 \times 573}v^2\right)$$

$$= 1.39 \times 10^{-8} v^2 \exp(-3.36 \times 10^{-6}\ v^2).$$

We can estimate the fraction of the molecules with speed greater than 2454 ms^{-1}.

Using the trapezium rule (or any numerical techniques) with ordinates at 2453, 2453 + 500, 2453 + 1000. The values are as follows:

Velocity, v	Probability, $W(v)$
2453	1.373×10^{-10}
2953	2.256×10^{-14}
3453	6.518×10^{-19}

Using trapezium rule, the fraction of molecules with speed greater than 2453 ms^{-1} is given by:

$$\text{Fraction of molescules} = \frac{500}{2}[(1.373 \times 10^{-10}) + (2 \times 2.256 \times 10^{-14})$$

$$+ (6.518 \times 10^{-19})]$$

$$f = 3.43 \times 10^{-8}.$$

Thus the time needed for deposition is given by 0.7 μs/ (3.43 $\times 10^{-8}$) that is 20.4 s.

(e) For destructive interference, optical path difference $= 2d = \frac{\lambda'}{2}$

where $\lambda' = \frac{\lambda_{air}}{n}$ is the wavelength in the coating.

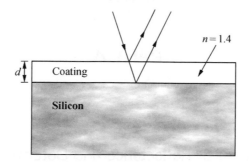

Fig. 3 – 10.

The relation is given by:

$$d = \frac{\lambda_{air}}{4n}.$$

Plugging in the given values, one gets $d = 105$ or 105.2 nm.

Experimental Competition

May 10, 2002

PART A

Time Available: $2\frac{1}{2}$ hours for Part A

Measurement of Reflectance and Determination of Refractive Index

Ⅰ. Background

The aim of this experiment is to measure the angular dependence of reflection of polarized light and to determine the refractive index of a semiconductor wafer.

When light falls on a semiconductor surface it will be partially reflected, partially transmitted and partially absorbed. The relative amount of light power reflected is called reflectance R, which is defined as the ratio of the reflected power I_r over the incident power I_i:

$$R = \frac{I_r}{I_i}. \tag{1}$$

The incident light may be resolved into two polarized components. One component is polarized parallel (labeled as p-polarization) to the plane of incidence, and the other polarized perpendicular (labeled as s-polarization) to the plane of incidence. For the red laser wavelength used in this study, the effects of absorption at the semiconductor surface are negligible. Under such condition, for an incident light from air onto a material the reflectance R_p and R_s, respectively for the p and s components, are given by the Fresnel equations:

$$\pm\sqrt{R_p} = \frac{n\cos\theta_i - \cos\theta_t}{\cos\theta_t + n\cos\theta_i}, \tag{2}$$

$$\pm \sqrt{R_s} = \frac{\cos \theta_i - n\cos \theta_t}{\cos \theta_i + n\cos \theta_t}, \tag{3}$$

where n is the refractive index of the material, θ_i is the angle of incidence, θ_r is the angle of reflection, and θ_t is the angle of transmission (or refraction), as shown in Fig. 3 – 11.

Direct measurements of R_s and R_p with $\theta_i = 0$ are practically not feasible. However, the Fresnel equations allow

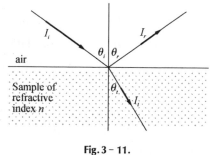

Fig. 3 – 11.

the calculation of n from R_s and R_p obtained for any oblique incident angle. A possible schematic diagram for the measurement is shown in Fig. 3 – 12, and a photograph of the suggested setup is shown in Fig. 3 – 13.

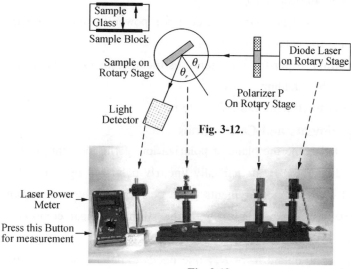

Fig. 3-12.

Fig. 3-13.

Note: The reference line for angular setting is on the upper left corner of the rotary stage.

II. Equipment and Materials

(1) A diode laser mounted on rotary stage together with 3V battery supply.

The laser emits a wavelength 650 nm.

Caution: Do not look into the laser beam. Watch out for strayed or scattered laser light.

Keep the glass and sample surfaces clean with tissues provided.

(2) A linear polarizer of diameter 20 mm mounted on a rotary stage.

Note: The angular setting of $0°$ of the rotary stage has no bearing on the polarization axis of the polarizer.

(3) A glass plate (refractive index 1. 57 for light of wavelength 650 nm) and a semiconductor wafer, fixed on the opposite sides of a rectangular sample block.

(4) Rotary optical stage with clamp for the sample block.

(5) A digital laser power meter, with light detector head mounted on a stand that can be revolved about the sample.

(6) One optical bench.

(7) Graph paper for reflected laser power vs angular setting of the rotary stage for polarizer (2 sheets).

(8) Graph paper for reflectance vs incident angle (4 sheets).

(9) Light-shielding board.

(10) A torch light and a flexible ruler.

Ⅲ. Experiments and Calculations

(1) Determine the plane of polarization of the incident laser light.

The diode laser emits partially linearly polarized light at 650 nm. For best results in your measurements to be performed, the polarization axis of the polarizer should be aligned with the strongest linear component of the laser light.

In order to obtain R_s or R_p, one needs to determine the orientation of the polarization axis of the polarizer in order to produce polarized light parallel or perpendicular to the plane of incidence. The axis of polarization of the polarizer can be inferred from the laser power reflected from the glass sample of known refractive index 1. 57.

With the optics aligned as accurately as possible,

(a) Determine the relative orientation of diode laser and polarizer (difference in degree between the angular settings of the diode laser rotary stage and the polarizer rotary stage) such that the polarizer is aligned with the strongest linear component of the laser light. In the measurements that follow, treat the polarizer and the source as a single system, rotating both together as necessary.

(b) Mount the glass sample on the rotary stage at the Brewster angle of incidence. Measure and plot the reflected laser power vs the angular setting (in degrees) of the polarizer. Hence determine the orientation of the polarization axis of the polarizer.

Note: You will have to press the button (as indicated in Fig. 3 – 13) on the laser power meter each time to take a reading.

(2) Measure the reflectance R_p and R_s of the semiconductor wafer.

Mount the sample block on the rotary stage so that the reflecting plane of the semiconductor wafer can be rotated about a vertical axis on the path of the incident light.

With the optics aligned as accurately as possible.

(a) Set the orientation of the incident light onto the semiconductor wafer such that it is polarized parallel to the plane of incidence.

Measure the reflected laser power and plot the values of R_p as a function of incident angle for a widest range of incident angles permitted by the experimental setup.

(b) Change the orientation of the incident light onto the semiconductor wafer such that it is polarized perpendicular to the plane of incidence.

Measure the reflected laser power and plot the values of R_s as a function of incident angle for a widest range of incident angles permitted by the experimental setup.

(3) Calculate the refractive index of the semiconductor wafer.

(a) From the Fresnel equations, show that $n = \sqrt{\dfrac{(1 \pm \sqrt{R_p})(1 + \sqrt{R_s})}{(1 \mp \sqrt{R_p})(1 - \sqrt{R_s})}}$.

From your graphs or otherwise determine the ranges of the angle of incidence where the signs of $\pm \sqrt{R_p}$ are positive and negative.

(b) Using the graphs obtained in Question (2), obtain six sets of values for R_p and R_s at angles of incidence of $20°$, $30°$, $40°$, $50°$, $60°$ and $80°$.

Calculate six values of the refractive index n of the semiconductor wafer using these six sets of values. Compute the mean value for n and estimate its standard deviation.

(c) Using the graphs obtained in Question (2), determine R_s and R_p at normal incidence by extrapolation. Hence calculate the average refractive index n of the semiconductor wafer from the results of extrapolation.

🔑 Answer

(1) Determine the plane of polarization of the incident laser light.

(a) Relative orientation of diode laser and polarizer (difference in degree between the angular settings of the diode laser rotary stage and the polarizer rotary stage).

The difference in the two angles (in degree) $= 0°$ or $180°$.

(b) Brewster angle of incidence θ_i for the glass plate $= \arctan 1.57 = 57.5°$.

Graph of reflected laser power versus angular setting of the polarizer rotary stage

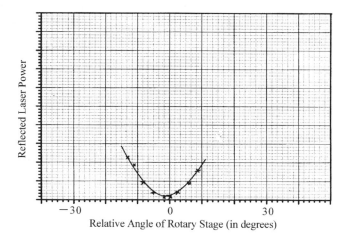

Fig. 3 – 14.

Angular setting of the polarizer rotary stage for axis of polarization parallel to the plane of incidence $= 30°$ or $210°$.

(The relative $0°$ of graph corresponds to $30°$ or $210°$.)

(2) Measure the reflectance R_p and R_s of the semiconductor wafer.

(a) Reflectance R_p (plane of polarization parallel to the plane of incidence)

Parameters of your instrument:

(i) Angular setting of the polarizer rotary stage $= 30°$ or $210°$.

(ii) Angular setting of the diode laser rotary stage $= 30°$ or $210°$.

(iii) Measured incidence laser power $I_i > 1.2(\text{mW})$.

Measurement, calculation of R_p and graph of R_p versus incident angle (see below).

(b) Reflectance R_s (plane of polarization perpendicular to the plane of incidence)

Parameters of your instrument:

(i) Angular setting of the polarizer rotary stage $= 120°$ or $300°$.

(ii) Angular setting of the diode laser rotary stage $= 120°$ or $300°$.

(iii) Measured incidence laser power $I_i > 1.2(\text{mW})$.

Measurement, calculation of R_s and graph of R_s versu incident angle (see below).

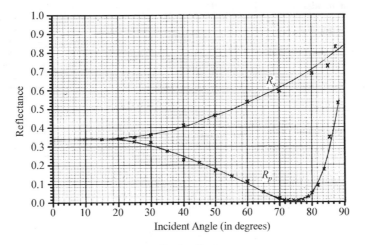

Fig. 3 - 15.

(3) Calculate the refractive index of the semiconductor sample.

(c) Equation relating the refractive index n to $\pm \sqrt{R_p}$ and $\sqrt{R_s}$.

Solve the simultaneous equations (2) and (3):

$$\pm \sqrt{R_p} = \frac{n\cos \theta_i - \cos \theta_t}{\cos \theta_t + n\cos \theta_i},$$

$$-\sqrt{R_s} = \frac{\cos \theta_i - n\cos \theta_t}{\cos \theta_i + n\cos \theta_t},$$

$$1 \pm \sqrt{R_p} = \frac{2n\cos \theta_i}{\cos \theta_t + n\cos \theta_i}, \quad 1 \mp \sqrt{R_p} = \frac{2\cos \theta_t}{\cos \theta_t + n\cos \theta_i},$$

$$1 + \sqrt{R_s} = \frac{2n\cos \theta_t}{\cos \theta_i + n\cos \theta_t}, \quad 1 - \sqrt{R_s} = \frac{2\cos \theta_i}{\cos \theta_i + n\cos \theta_t},$$

$$n = \sqrt{\frac{(1 \pm \sqrt{R_p})(1 + \sqrt{R_s})}{(1 \mp \sqrt{R_p})(1 - \sqrt{R_s})}}.$$

The sign(s) of $\pm \sqrt{R_p}$: $+ \sqrt{R_p}$ angle of incidence less than Brewster angle;

$- \sqrt{R_p}$ angle of incidence larger than Brewster angle.

(d) Six values of n

angle of incidence θ_i	R_p	R_s	n
20	0.340	0.340	3.80
30	0.325	0.367	3.86
40	0.225	0.417	3.61
50	0.175	0.467	3.60
60	0.108	0.540	3.60
80	0.025	0.780	3.45

Mean and standard deviation σ_n for n

$$n \text{ (mean)} = 3.65$$

$$\sigma_n = \sqrt{\frac{\sum (n_i - \bar{n})^2}{n}}.$$

(σ_n is of the order of $0.15 \sim 0.20$.)

Note: The value of n for the glancing angle of 80° may be excluded in the calculation of σ_n for better accuracy. The calculation above includes the glancing angle.

(e) Reflectance R_s and R_p of the semiconductor at normal incidence.

$$R_s = \sim 34\%,$$
$$R_p = \sim 34\%.$$

At normal incidence, $R = R_s = R_p$, from Fresnel equations.

Take average value of R_s and R_p.

Refractive index $n = \dfrac{1+\sqrt{R}}{1-\sqrt{R}} = 3.80.$

PART B

Time Available: $2\frac{1}{2}$ hours for Part B

Objective

To study how the frequency of vibration of a tuning fork varies with an equal mass clamped on each of its prongs (at a definite point near the prong tip), and hence to determine the pair of unknown masses X similarly attached to the prongs.

The Stroboscope

The experiment will make use of a stroboscope (strobe) which is a simple electronic device consisting of a discharge lamp which can be made to flash for a short duration with a high intensity at highly regular intervals. The strobe enables the frequency of a rotating or vibrating object to be measured without the need for any direct physical contact with the moving object.

Caution: The strobe has a finite life time, specified in maximum number of flashes obtainable. Do not leave it running idly when you are not using it.

Consider a particle rotating with uniform circular motion being illuminated by a strobe. If the flash frequency is a multiple or sub-multiple of that of the motion, the particle will appear stationary. It follows that the periodicity of the circular motion of the particle can be determined by tuning the frequency of the light flash.

Suppose the frequency of rotation of the particle is x Hz, and that of flashing is y Hz. Then, in the time interval of $\dfrac{1}{y}$ s between two successive flashes the particle would have moved through an angle $\dfrac{2\pi x}{y}$.

If $\dfrac{y}{x}$ is an irrational number so that it cannot be expressed as a ratio of two integers, then the particle would not appear stationary but would appear to rotate slowly in the forward or backward direction depending on whether $\dfrac{y}{x}$ is just slightly smaller or larger than some rational number nearby.

If $\dfrac{y}{x} = \dfrac{q}{p}$ where p and q are integers, then the strobe would flash q times for every p complete cycles. Furthermore, if p and q have no common factors between them (assumed throughout this write-up), then each flash would show a different position of the particle. Thus the particle will exhibit q stationary positions under the strobe flashlight.

If q becomes too large, it might be difficult to count the number of stationary positions displayed by the rotating particle.

The above theory applied to the rotating particle can be similarly applied to that of a tuning fork vibrating in simple harmonic motion if we regard the vibrational motion as equivalent to the motion of the projection of the rotating particle's position on a given diameter of the circle of motion. However, in this case, because the vibrating object retraces the same path in the opposite direction every half cycle, there is a chance, though very remote, that an image in one half of a vibration cycle coincides with that in the next half cycle. It would result in only one image (but of double the intensity) being recorded, instead of two. This freak coincidence should be guarded against in an experimental observation.

Identification of Fundamental Synchronism

Fundamental synchronism is obtained when the lamp flashes once for every cycle of rotation or vibration of the mechanism under observation, so that the object appears to stop at one stationary position. However, it will be appreciated that a similar and indistinguishable result will also occur when the flash frequency is a sub-multiple $\left(\frac{1}{2}, \frac{1}{3}, \frac{1}{4}, \text{etc}\right)$ of the object movement frequency. Thus if the object movement frequency is totally unknown, when adjusting for fundamental synchronism, a *safe* procedure is to start at a high flash frequency, when multiple images are obtained, and then slowly reduce the flash frequency until *the first* single image appears. This procedure should be adopted in all measurements to check for fundamental synchronism.

Multiples of Fundamental Frequency

Multiples of fundamental frequency occur when the strobe is flashing at a higher rate than the cyclical frequency of the object under observation. The converse when the strobe flashing rate is lower than that of the moving object is referred to as sub-multiples of fundamental frequency.

If the lamp is flashed at a frequency q times the rotational frequency of the particle, multiple images can be seen. In such a situation, a rotating particle will appear as several stationary images spaced equally around the circumference. Twice this frequency, or $\frac{q}{p} = 2$, will produce two such images at π radians apart, and three times this frequency, or $\frac{q}{p} = 3$, will yield three images at $\frac{2\pi}{3}$ radians spacing, etc. The particle rotational frequency is then given by the flash frequency divided by the number of images seen. In general, if $q > p > 1$, then the strobe would flash q times for every p cycles of the particle motion, and so there will still be q stationary positions.

Sub-multiples of Fundamental Frequency

Here, $\frac{q}{p}$ is less than one. If the strobe frequency is exactly $\frac{1}{p}$ times that of the object movement, where $p > 1$, then the object would have moved through p cycles for every flash, and only one stationary image is seen. If $p > q > 1$, then the strobe would have flashed q times for every p full cycles of the object movement, and the number of stationary images seen would be q.

The Tuning Fork

A tuning fork is designed to vibrate at a fundamental frequency with no harmonics after it is struck. The two prongs of the fork are symmetrical in every respect so that they move in perfect anti-phase and exert, at any instant, equal and opposite forces on the central holder. The net force on the holder is therefore always zero so that the holder does not vibrate, and hence holding it firmly will not cause any undesirable damping. For the same reason the prongs of a tuning fork cannot vibrate in like phase as this will result in a finite oscillatory force on its holder which would cause the vibration to dampen away very quickly.

It is possible to lower the fundamental frequency of the tuning fork by loading an equal weight on each arm. The loading on the arms has to be symmetrical in order to minimise damping of vibration.

For such a loaded tuning fork, the period T of vibration is given by:

$$T^2 = A(m+B),$$

where A is a constant depending on the size, shape and mechanical properties of the tuning fork material and B is a constant depending on the effective mass of each vibrating arm.

Items of Apparatus provided:

1. A stroboscope with digital readout.

2. A mini-torch light.

3. A tuning fork with a 31.6g weight loaded symmetrically on each prong and with the centre-of-mass of the weight coinciding with the point P marked clearly on each prong.

4. Two paper clamps with two detachable levers. The levers are used only to open the clamps, and they should be removed when you are doing the experiment.

5. A pair of equal unknown masses X.

6. A series of the following known masses (in pairs): 5 g, 10 g, 15 g, 20 g, 25 g.

7. Regular graph papers (5 sheets).

Experimental Steps

Step 1. Fundamental synchronism and measurements of multiple frequencies.

(a) Obtain fundamental synchronism between the strobe flash and the vibrating tuning fork loaded with the original 31. 6g mass on each prong. By dislodging the mass temporarily, check to make sure that the mass is pre-clamped with its centre-of-mass located at the point P (which is marked on the prong but hidden by the mass). Record its fundamental flash frequency.

(b) Keeping the flash frequency above the fundamental frequency, try to discover *as many readings of flash frequencies* as possible which yield observable stationary images of the (31. 6g-loaded) tuning fork frequency. Identify their different $\frac{q}{p}$ values.

(c) Tabulate your data (in the order of increasing $\frac{q}{p}$) as follow, keeping $\frac{q}{p}$ as a rational fraction:

Strobe Reading	Number of Stationary Images	$\frac{q}{p}$ value

Plot a straight-line graph of all the observed strobe flash frequencies against the corresponding multiple of the tuning fork frequency. Identify each data point on the graph with its $\frac{q}{p}$ value.

Step 2. Measurements of sub-multiple frequencies.

(a) Keeping the strobe frequency below the fundamental frequency of the (31. 6g-loaded) tuning fork, obtain readings of *all observable strobe frequencies which yield stationary images.*

(b) Tabulate your readings as in question 1, but in the order of decreasing $\frac{q}{p}$, and plot a straight-line graph of all the observed strobe frequencies against the corresponding sub-multiple of the (31. 6g – loaded) tuning fork frequency. Identify each data point on the graph with its $\frac{q}{p}$ value.

Step 3. Determination of the pair of unknown masses X.

(a) Remove the 31. 6g loading mass from each prong (which would also reveal the point P marked on the prong) and obtain the resulting vibrational frequency of the unloaded tuning fork.

(b) Next, obtain the vibrational frequencies of the tuning fork with each prong loaded with known masses m of 5 g, 10 g, 15 g, 20 g and 25 g respectively. Ensure that in each case the centre-of-mass of the load coincides with the point P. Note that the value of m as labelled on the mass is the total mass of both the mass itself and that of the given paper clamp (with both its levers removed) used to clamp it.

(c) Tabulate your results using your data obtained in (b) and plot a graph of T^2 against m. Obtain the slope, and the intercept on the m-axis.

(d) Replace the known loading masses with the unknown masses X, and obtain the vibrational frequency under this loading. Deduce X. Again, note that X also includes the mass of the paper clamp (with both its levers removed).

Solution

Step 1. Fundamental Synchronism and Multiple Frequencies

Strobe reading (Hz)	No. of stationary images	$\frac{q}{p}$ value
65. 1	1	1
81. 7	5	$1\frac{1}{4}$

Cont.

Strobe reading (Hz)	No. of stationary images	$\dfrac{q}{p}$ value
87. 2	4	$1\dfrac{1}{3}$
98. 1	3	$1\dfrac{1}{2}$
109. 0	5	$1\dfrac{2}{3}$
130. 8	2	2
163. 5	5	$2\dfrac{1}{2}$
196. 2	3	3
261. 4	4	4

Multiples of Fundamental Freq.

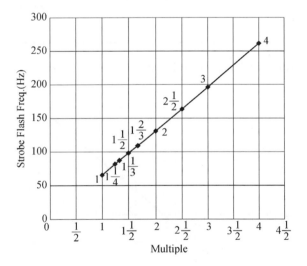

Fig. 3 - 16.

Step 2: Sub-multiple frequencies

Strobe reading (Hz)	No. of stationary images	$\dfrac{q}{p}$ value
65. 1	1	1
49. 0	3	$\dfrac{3}{4}$

Cont.

Strobe reading (Hz)	No. of stationary images	$\frac{q}{p}$ value
43. 6	2	$\frac{2}{3}$
39. 2	3	$\frac{3}{5}$
32. 7	1	$\frac{1}{2}$
26. 2	2	$\frac{2}{5}$
21. 8	1	$\frac{1}{3}$
16. 3	1	$\frac{1}{4}$
13. 0	1	$\frac{1}{5}$

Sub-Multiples of Fundamental Freq.

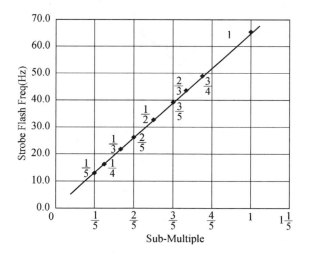

Fig. 3 – 17.

Step 3: Determination of *X*

Weight (g)	Frequency (Hz)	Period (*T*) (ms)	T^2 (μs^2)
0	128. 3	7. 794	60. 8
5	106. 6	9. 381	88. 0
10	94. 0	10. 64	113
12. 8	87. 5	11. 43	131
15	83. 2	12. 02	144
20	76. 0	13. 16	173
25	70. 1	14. 27	204
31. 6	65. 1	15. 36	236

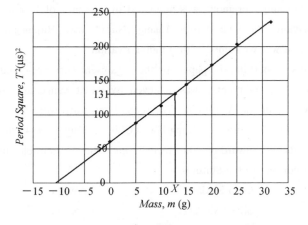

Fig. 3 - 18.

Intercept on *m*-axis $=-10. 5$ g.

Best fit slope $= 230/40. 5 = 5. 7\ \mu s^2/g$.

Value of $X = 12. 8$ g.

Minutes of the Fourth Asian Physics Olympiad

Bangkok (Thailand), April 20 – 29, 2003

1. The following 10 countries were present at the 4th Asian Physics Olympiad:
Australia (8 students + 2 leaders), Indonesia (8 students + 2 leaders), Israel (8 students + 2 leaders), Kyrgyzstan (5 students + 1 leader), Laos (8 students + 2 leaders), Pakistan (2 students + 1 leader), Philippines (1 student + 1 leader), Chinese Taipei (8 students + 2 leaders + 1 observer), Thailand (8 students + 2 leaders + 1 observer) and Vietnam (8 students + 2 leaders + 10 observers).

This year Laos and Pakistan participated in the APhO for the first time.

2. The Opening Ceremony was honoured by presence of Her Royal Highness Princess Galyani Vadhana Krom Luang Naradhiwas Rajanagariondra. The Respected Guest was welcomed with standing ovation.

3. Results of marking the papers by the organisers were presented:
The best score (45. 90 points) was achieved by Pawit Sangchant from Thailand (Absolute Winner of the IV APhO). The second and third were Widagdo Setiawan (Indonesia) — 44. 60 points and Kang-Hao (Chinese Taipei) — 44. 10 points. The following limits for awarding the medals and the honourable mention were established according to the Statutes:

Gold Medal:	40 points,
Silver Medal:	34 points,
Bronze Medal:	29 points,
Honourable Mention:	22 points.

According to the above limits 14 Gold Medals, 13 Silver Medals, 11 Bronze Medals and 12 Honourable Mentions were awarded. The list of the scores of the winners and the students awarded with honourable mentions were distributed to all the delegations.

4. In addition to the regular prizes a number of special prizes were awarded:
- for the Absolute Winner: Pawit Sangchant (Thailand)
- for the most creative solution in the theoretical part of the competition: Tse-Yu Chen (Chinese Taipei)
- for the most creative solution in the experimental part of the competition: Rangga Perdana Budoyo (Indonesia)

- for the best score among female participants: Veronika Ulychny (Israel)
- for the best result among the "new" countries: Sara Ijaz Gilani (Pakistan)

5. Dr. Colin Taylor (Australia) and Dr. Waldemar Gorzkowski (Honorary President of the APhOs), acting on behalf of all the Members of the International Board, expressed deep thanks to Dr. Sirikorn Maneerin, Deputy Minister of Education and Chairman of the Advisory Committee, Professor Sakda Siripant, Chairman of the Organising Committee, Dr. Wijit Senghapant, Chairman of the International Board Committee, Dr. Wudhibhan Prachyabrued, Chairman for the Examination Papers, and Associate Professor Suwan Kusamran, Executive Secretary of the Organising Committee, for excellent organisation and execution of the Ⅳ APhO.

6. The Israeli delegation provided the Permanent Secretariat of the APhOs with the letter from the Israeli Ministry of Education. According to the letter, the APhO'2011 will be organized in Israel. Thus, the last version of the list of the organizers of the next APhOs is:

V	2004	Hanoi	Vietnam	April 26 – May 4(9)
Ⅵ	2005	Lingga Island, Riau	Indonesia	confirmed orally
Ⅶ	2006	not decided yet	Georgia	not reconfirmed
Ⅷ	2007	not decided yet	China	preliminary contacts
Ⅸ	2008	not decided yet	Australia	preliminary contacts
Ⅹ	2009	not decided yet	Uzbekistan	preliminary contacts
Ⅺ	2010	not decided yet	Malaysia	preliminary contacts
Ⅻ	2011	not decided yet	Israel	confirmed

The Permanent Secretariat will ask the Georgian leaders (not present this year) for written reconfirming willingness of Georgia (expressed orally last year) to organise the competition in 2006 by end of July 2003.

7. Acting on behalf of the organisers of the next Asian Physics Olympiad Prof. Phan Hong Khoi announced that the V Asian Physics Olympiad will be organised in Hanoi (Vietnam) from April 26th to May 4th, 2004 and cordially invited all the participating countries to attend the competition.

Assoc. Prof. Suwan Kusamran	**Prof. Ming-Juey Lin**	**Dr. Yohanes Surya**
Executive Secretary of the	Secretary of the APhOs	President of the APhOs
Ⅳ APhOs		

Bangkok, 30.04.2003

Theoretical Competition

April 23, 2003   Time available: 5 hours

Problem 1
Satellite's Orbit Transfer

In the near future we ourselves may take part in launching of a satellite which, in point of view of physics, requires only the use of simple mechanics.

(a) A satellite of mass m is presently circling the Earth of mass M in a circular orbit of radius R_0. What is the speed (u_0) of mass m in terms of M, R_0 and the universal gravitation constant G?

(b) We are to put this satellite into a trajectory that will take it to point P at distance R_1 from the centre of the Earth by increasing (almost instantaneously) its velocity at point Q from u_0 to u_1. What is the value of u_1 in terms of u_0, R_0, R_1?

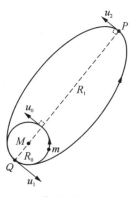

Fig. 4 - 1.

(c) Deduce the minimum value of u_1 in term of u_0 that will allow the satellite to leave the Earth's influence completely.

(d) (Referring to part (b).) What is the velocity (u_2) of the satellite at point P in terms of u_0, R_0, R_1?

(e) Now, we want to change the orbit of the satellite at point P into a circular orbit of radius R_1 by raising the value of u_2 (almost instantaneously) to u_3.

What is the magnitude of u_3 in terms of u_2, R_0, R_1?

(f)

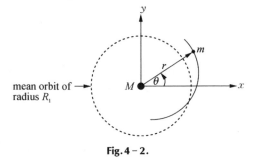

mean orbit of radius R_1

Fig. 4 - 2.

If the satellite is slightly and instantaneously perturbed in the radial direction so that it deviates from its previously perfectly circular orbit of radius R_1, derive the period of its oscillation T of r about the mean distance R_1.

Hint: Students may make use (if necessary) of the equation of motion of a satellite in orbit:

$$m\left[\frac{d^2}{dt^2}r - \left(\frac{d}{dt}\theta\right)^2 r\right] = -G\frac{Mm}{r^2},\tag{1}$$

and the conservation of angular momentum:

$$mr^2\frac{d}{dt}\theta = \text{constant.}\tag{2}$$

(g) Give a rough sketch of the whole perturbed orbit together with the unperturbed one.

Solution

(a) $\dfrac{mu_0^2}{R_0} = \dfrac{GMm}{R_0^2}$, $u_0 = \sqrt{\dfrac{GM}{R_0}}$.

(b) Conservation of angular momentum: $mu_1 R_0 = mu_2 R_1$.

Conservation of energy: $\dfrac{1}{2}mu_2^2 - \dfrac{GMm}{R_1} = \dfrac{1}{2}mu_1^2 - \dfrac{GMm}{R_0}$

$$\left[\left(\frac{R_0}{R_1}\right)^2 - 1\right]u_1^2 = 2GM\left(\frac{1}{R_1} - \frac{1}{R_0}\right)$$

$$\frac{(R_0 - R_1)(R_0 + R_1)}{R_1^2}u_1^2 = 2GM\frac{R_0 - R_1}{R_0 R_1}$$

$$u_1 = \sqrt{\frac{GM}{R_0}}\sqrt{\frac{2R_1}{R_1 + R_0}} = u_0\sqrt{\frac{2R_1}{R_1 + R_0}}.$$

(c) $\displaystyle\lim_{R_1 \to \infty} u_1 = \sqrt{2}\,u_0$.

(d) $u_2 = u_1\dfrac{R_0}{R_1} = u_0\dfrac{\sqrt{2}R_0}{\sqrt{R_1(R_1 + R_0)}}$.

(e) $u_3 = \sqrt{\dfrac{GM}{R_1}} = \sqrt{\dfrac{GM}{R_0}}\sqrt{\dfrac{R_0}{R_1}} = u_0\sqrt{\dfrac{R_0}{R_1}}$

$$= \sqrt{\frac{R_0}{R_1}} \sqrt{\frac{R_1 (R_1 + R_0)}{\sqrt{2} R_0}} u_2.$$

$$u_3 = u_2 \sqrt{\frac{R_1 + R_0}{2R_0}}.$$

(f) Combining equations (1) and (2):

$$\frac{d^2}{dt^2} r - \frac{C}{mr^3} = -\frac{GM}{r^2},$$

and for the circular orbit of radius R_1 we have $\dfrac{C}{m} = GMR_1$,

hence

$$\frac{d^2}{dt^2} r - \frac{GMR_1}{r^3} = -\frac{GM}{r^2},$$

putting $r = R_1 + \eta$, where $\eta \ll R_1$

$\therefore$

$$\frac{d^2}{dt^2}\eta - \frac{GMR_1}{R_1^3 \left(1 + \dfrac{\eta}{R_1}\right)^3} = -\frac{GM}{R_1^2 \left(1 + \dfrac{\eta}{R_1}\right)^2}$$

$$\frac{d^2}{dt^2}\eta - \frac{GM}{R_1^2}\left(1 - 3\frac{\eta}{R_1}\right) \approx -\frac{GM}{R_1^2}\left(1 - 2\frac{\eta}{R_1}\right)$$

$$\frac{d^2}{dt^2}\eta \approx -\frac{GM}{R_1^3}\eta \ .$$

The frequency of oscillation about mean distance is $f = \dfrac{1}{2\pi}\sqrt{\dfrac{GM}{R_1^3}}$.

The period $T = \dfrac{1}{f} = 2\pi\sqrt{\dfrac{R_1^3}{GM}}$.

Note that this period is the same as the orbital period.

(g)

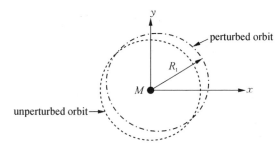

Fig. 4 – 3.

Problem 2

Optical Gyroscope

In 1913 Georges Sagnac (1869 – 1926) considered the use of a ring resonator to search for the aether drift relative to a rotating frame. However, as often happen, his results turned out to be useful ways that Sagnac himself never dreamt of. One of those applications is the Fibre – Optic Gyroscope (FOG) which is based upon a simple phenomena, first observed by Sagnac. The essential physics associated with the Sagnac effect is due to the phase shift caused by two coherent beams of light being sent around a rotating ring of optical fibre in the opposite directions. This phase shift is also used to determine the angular speed of the ring.

As shown in a schematic diagram in Fig. 4 – 4, a light wave enters a circular optical fibre light path of radius R at point P on the rotating platform with a uniform angular speed Ω, in the clockwise direction. Here the light wave is split into two waves which travels in the opposite directions, clockwise (CW) and counter clockwise (CCW), through the ring. The refractive index of optical fibre material is μ.

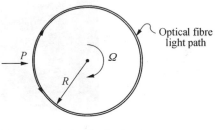

Fig. 4 – 4.

(a) Practically, the orbital speed of the ring is much less than the speed of light such that $(R\Omega)^2 \ll c^2$, find the time difference $\Delta t = t^+ - t^-$ where t^+ and t^- denote the round-trip transit time of the CW and CCW beam respectively. Give your answer in term of area A enclosed by the ring.

(b) Find the path difference, ΔL, for the CW and CCW beams to complete one round-trip of the rotating ring.

(c) For a circular fibre-optic of radius, $R = 1$ m, what is the maximum value of ΔL for the rotation of the earth?

(d) In part (c), the measurement could be amplified by increasing number of turns in fibre-optic coil, N, find the phase difference, $\Delta\theta$, for lights to complete the turns.

The second scheme of the Optical Gyroscope is Ring Laser Gyroscope (RLG). This could be accomplished by the inclusion of active laser cavity into an equilateral triangular ring, of length L, as illustrated in Fig. 4 – 5. The laser source here will generate two amplified coherent light sources propagating in the opposite directions. In order to sustain the laser oscillation in this triangular ring resonator, the perimeter of the ring must be equal to the integer multiple of wavelength λ. Etalon, additional component inserted into the ring, is possible to cause frequency selective losses in the ring resonator, so that the undesired modes can be damped and suppressed.

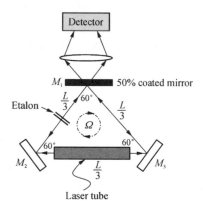

Fig. 4 – 5. Schematic illustration of the Ring Laser Gyroscope.

Fig. 4 – 6. Illustration of the Ring Laser Gyroscope discussed in this problem.

(e) Find the time difference of the transit in clockwise and counterclockwise, Δt, for the case of the triangular ring as shown in Fig. 4 – 5. Give the answer in terms of Ω and the area A enclosed by the ring. Show that this result is the same as that of the circular ring.

(f) If the ring is rotating with an angular frequency Ω as shown in

Fig. 4 – 5, there will be frequency difference between CW and CCW measurements. What is the observed beat frequency, Δv, between the CW and CCW beams in terms of L, Ω, λ?

Solution

The light wave moves with speed $c' = \dfrac{c}{\mu}$ in the medium having refractive index μ.

Wavelength of light in medium $\lambda' = \dfrac{\lambda}{\mu}$, where λ is the wavelength of light in vacuum.

(a) Transit time for the CW beam:

$$t^+ = \frac{2\pi R + R\Omega t^+}{c'} = \frac{2\pi R}{c'}\left(1 - \frac{R\Omega}{c'}\right)^{-1};$$

transit time for the CCW beam:

$$t^- = \frac{2\pi R - R\Omega t^-}{c'} = \frac{2\pi R}{c'}\left(1 + \frac{R\Omega}{c'}\right)^{-1};$$

the time difference between t^+ and t^- : $\Delta t = \dfrac{4\pi R^2 \Omega}{(c')^2 - R^2\Omega^2}$;

since $(R\Omega)^2 \ll c'^2$,

$$\Delta t \approx \frac{4\pi R^2 \Omega}{(c')^2}.$$

(b) The round-trip optical path difference, ΔL, is given by

$$\Delta L = c\Delta t = \left(\frac{4\pi R^2 \Omega}{c}\right)\mu^2.$$

(c) $\Delta L = 6.9 \times 10^{-12}$ m.

Indication that the gyroscope has to be placed at the north/south pole.

(d) The corresponding optical phase difference $\Delta\theta$ is

$$\Delta\theta = \frac{2\pi\Delta L}{\lambda} = \left(\frac{8\pi^2 R^2 \Omega}{c\lambda}\right)\mu, \text{ where } \lambda' = \frac{\lambda}{\mu}.$$

For N turns of fiber optic ring,

$$\Delta\theta = \frac{8\pi^2 R^2 N\Omega}{c'\lambda'} = \left(\frac{8\pi^2 R^2 N\Omega}{c\lambda}\right)\mu^2.$$

(e) The figure shows the triangular ring rotating about the centre O with the angular speed Ω in the clockwise direction. Without loosing generality, let us first consider the velocity of light along AC in the CW and CCW direction,

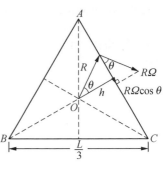

$$v_\pm = c \pm R\Omega\cos\theta = c \pm \Omega h,$$

where h is constant.

Fig. 4 – 7.

$$\tau_\pm = \frac{\dfrac{L}{3}}{v_\pm} = \frac{\dfrac{L}{3}}{c \pm \Omega h} \approx \frac{L}{c}\left(1 \mp \frac{\Omega h}{c}\right),$$

where $\tau_\pm$ is the time taken for light traveling along AC in the CW and CCW.

$$t_\pm = \frac{L}{v_\pm} = \frac{L}{c \pm \Omega h} \approx \frac{L}{c}\left(1 \mp \frac{\Omega h}{c}\right),$$

where L is the perimeter of the triangular ring.

Therefore, the time difference of light traveling in one complete cycle.

$$\Delta t = \frac{2\Omega L h}{c^2} = \frac{4\Omega}{c^2}\left(\frac{1}{2}Lh\right) = \frac{4\Omega A}{c^2},$$

where A is the area of the triangular ring.

(f) The resonance frequencies associated with $L_\pm$ corresponding to the effective cavity lengths seen by CW and CCW propagating beams respectively is

$$L_+ = ct^+ \approx L\left(1 - \frac{\Omega h}{c}\right),$$

$$L_- = ct^- \approx L\left(1 + \frac{\Omega h}{c}\right),$$

where $L_\pm$ is the perimeter of the equilateral triangle in the CW ($+$) and CCW ($-$) and we also use the fact that $h\Omega \ll c$.
Therefore,

$$\Delta L = L_- - L_+ = 2L \frac{\Omega h}{c}.$$

The condition to sustain the laser oscillation (given in the problem),

$$v_{\pm} = \frac{m}{L_{\pm}} c, \; m = 1, 2, 3, \ldots$$

$$\Delta v = v_- - v_+ = \frac{m}{L_-} c - \frac{m}{L_+} c \approx mc \frac{\Delta L}{L^2} = v \frac{\Delta L}{L}.$$

The approximation arises from $L_+ L_- \approx L^2$, where L is the perimeter of the triangular ring.

Hence,
$$\Delta v = \frac{\Delta L}{L} v = \frac{4A}{Lc} v \Omega = \frac{1}{3\sqrt{3}} \frac{L}{\lambda} \Omega.$$

Problem 3
Plasma Lens

The physics of the intense particle beams has a great impact not only on the basic research but also the applications in medicine and industry. The plasma lens is a device to provide an ultra-strong final focus at the end of the linear colliders. To appreciate the possibilities of the plasma lens it is quite natural to compare this with the usual magnetic and electrostatic lenses. In magnetic lenses, focusing capability is proportional to the magnetic field gradient. The practical upper limit of the quadrupole focusing lens is in the order of 10^2 T/m, while for plasma lens having density of 10^{17} cm^{-3}, its focusing capability is equivalent to a magnetic field gradient of 3×10^6 T/m (about four orders of magnitude more than that of a magnetic quadrupole lens).

In what follows, we will illuminate why the intense relativistic particle beams could produce self-focusing beams and do not blow themselves apart in free space.

(a) Consider a long cylindrical electron beam of uniform number density n and average speed v (both quantities in laboratory frame). Derive the expression for the electric field at a point at distance r from the central axis of the beam using classical electromagnetics.

(b) Derive the expression for the magnetic field at the same point as in a).

(c) What is then the net outward force on the electron in the electron beam passing that point?

(d) Assuming that the expression obtained in (c) is applicable at relativistic velocities, what will be the force on the electron as v approaches the speed of light c, where $c = \dfrac{1}{\sqrt{\varepsilon_0 \mu_0}}$?

(e) If an electron beam of radius R enters into a plasma of uniform density $n_0 < n$ (The plasma is an ionized gas of ions and electrons with equal charge density.), what will be the net force on the stationary plasma ion at distance r' outside the electron beam long after the beam entering the plasma. You may assume that the density of the plasma ions remains constant and the cylindrical symmetry is maintained.

(f) After long enough time, what is the net force on an electron at distance r from the central axis of the beam in the plasma, assuming $v \rightarrow c$ provided that the density of the plasma ions remains constant and the cylindrical symmetry is maintained?

Solution

(a) From Gauss's law, $\qquad E_r = -\dfrac{ner}{2\varepsilon_0}$.

(b) From Ampere's law, $\qquad B_\theta = -\dfrac{\mu_0 ner v}{2}$.

(c) The net Lorentz force is

$$F = \left(\frac{ne^2 r}{2\varepsilon_0} - \frac{\mu_0 ne^2 rv^2}{2} \right) r = \frac{ne^2 r}{2\varepsilon_0} \left(1 - \frac{v^2}{c^2} \right) r$$

where $\qquad c = \dfrac{1}{\sqrt{\varepsilon_0 \mu_0}}$.

(d) $F_r \rightarrow 0$ as $v \rightarrow c$, this implies the electric force and magnetic force cancel each other out.

(e) The stationary plasma particles have $v = 0$, hence $F_{r'} = \pm eE_{r'}$,

where
$$E_{r'} = -\frac{neR^2}{2\varepsilon_0 r'} + \frac{n_0 er'}{2\varepsilon_0},$$

for positive ion
$$F_{r'} = -\frac{ne^2 R^2}{2\varepsilon_0 r'} + \frac{n_0 e^2 r'}{2\varepsilon_0},$$

for electron
$$F_{r'} = \frac{ne^2 R^2}{2\varepsilon_0 r'} - \frac{n_0 e^2 r'}{2\varepsilon_0},$$

and there is no cancellation from the magnetic force.

As a result the plasma electrons will be blown out, and the ions are pulled in.

(f) The net force on the electron beam in plasma medium is given by,

$$F = \frac{ne^2 r}{2\varepsilon_0}\left(1 - \frac{v^2}{c^2}\right)r - \frac{n_0 e^2 r}{2\varepsilon_0}\, r,$$

in the limit $v \to c$, $F \approx -\dfrac{n_0 e^2 r}{2\varepsilon_0}\, r.$

Experimental Competition

April 25, 2003

I. Determination of Capacitance

Background

It is known that capacitors play a significant role in the electrical circuits. There are several methods of measurements of the capacitance of a capacitor. In this experiment you are required to perform the experiment in order to determine the capacitance of an AC capacitor using a simple electrical circuitry.

In Fig. 4 – 8, a capacitor of capacitance C and a resistor of resistance R are connected in series to the alternating voltage source of mains frequency. The electrical power which is dissipated at the resistor R depends on the values of ε_0, C, R and frequency of the mains f. Graphical analysis of this relationship can be used to determine C.

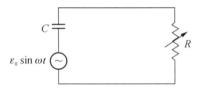

Fig. 4 – 8. AC Circuit for determination of capacitance C.

Materials and apparatus

1. Capacitor.

2. Three resistors of known values with $\pm 5\%$ errors ($R_A = 680\ \Omega$, $R_B = 1500\ \Omega$ and $R_C = 3300\ \Omega$) as shown in Fig. 4 – 9.

3. Step-down isolation transformer for alternating voltage source of $f = 50$ Hz.

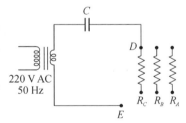

Fig. 4 – 9. A schematic diagram of the equipment used.

4. Digital voltmeter.

5. Electrical connectors.

6. Linear graph papers.

Warning: The digital multimeter in this experiment will be used for measuring the rms voltage ($\widetilde{V}$) across R only. Do not use it to measure in other modes.

Instructions

(a) Derive the expression for the average power dissipation $\overline{P}$ in resistor R in terms of ε_0, R, C and ω.

(b) Deduce the condition for which $\overline{P}$ is a maximum.

(c) Convert the dependence found in (a) into a linear dependence of certain quantities α and β.

(d) Measure the root mean square (effective) voltage V across R for each of all possible combinations of R_A, R_B and R_C.

(e) Plot $\overline{P}$ versus R and from this graph compute the value of capacitance C.

(f) From (c), draw the graph of α versus β and determine capacitance C.

(g) Estimate the uncertainties in the values of C obtained in (e) and (f).

Solution

(a) $\displaystyle \overline{P} = I^2 R = \frac{\frac{1}{2}\varepsilon_0^2}{R^2 + \left(\frac{1}{\omega C}\right)^2} R.$

(b) $\displaystyle \frac{\mathrm{d}}{\mathrm{d}R}\overline{P} = 0,$

$$\frac{\mathrm{d}}{\mathrm{d}R}\overline{P} = \frac{\mathrm{d}}{\mathrm{d}R} \frac{\frac{1}{2}\varepsilon_0^2}{R^2 + \left(\frac{1}{\omega C}\right)^2} R$$

$$= \frac{1}{2}\varepsilon_0^2 \frac{R^2 + \left(\frac{1}{\omega C}\right)^2 - R(2R)}{\left[R^2 + \left(\frac{1}{\omega C}\right)^2\right]^2},$$

condition for $\overline{P}_{\max}$: $\displaystyle R = \frac{1}{\omega C}.$

(c) $\displaystyle \overline{P} = \frac{\frac{1}{2}\varepsilon_0^2}{R^2 + \left(\frac{1}{\omega C}\right)^2} R = \frac{\frac{1}{2}\varepsilon_0^2 R}{R^2\left[1 + \left(\frac{1}{R\omega C}\right)^2\right]}$

$$= \frac{\frac{1}{2}\varepsilon_0^2}{R\left[1+\left(\frac{1}{R\omega C}\right)^2\right]}$$

$$\Rightarrow \frac{1}{R\overline{P}} = \frac{2}{\varepsilon_0^2}\left(1+\frac{1}{R^2}\frac{1}{\omega^2 C^2}\right)$$

$$\frac{1}{R\overline{P}} = \frac{1}{V^2} = \frac{2}{\varepsilon_0^2} + \frac{2}{\varepsilon_0^2}\left(\frac{1}{\omega C}\right)^2\frac{1}{R^2}.$$

Note: The linear graph will be $\dfrac{1}{R\overline{P}}$ or $\dfrac{1}{V^2}$ versus $\dfrac{1}{R^2}$. If a is the slope and

b is the intercept with the Y axis, then $\dfrac{1}{\omega^2 C^2} = \dfrac{a}{b} \Rightarrow C = \dfrac{1}{\omega}\sqrt{\dfrac{b}{a}}$.

An alternative method:

$$\frac{V^2}{R^2} = \frac{\frac{1}{2}\varepsilon_0^2}{R^2 + \left(\frac{1}{\omega C}\right)^2},$$

$$\frac{R^2}{V^2} = \left[R^2 + \left(\frac{1}{\omega C}\right)^2\right]\frac{2}{\varepsilon_0^2},$$

$$\frac{1}{V^2} = \left[1 + \left(\frac{1}{\omega C}\right)^2\frac{1}{R^2}\right]\frac{2}{\varepsilon_0^2},$$

$$\frac{1}{V^2} = \frac{2}{\varepsilon_0^2} + \frac{2}{\varepsilon_0^2}\left(\frac{1}{\omega C}\right)^2\frac{1}{R^2},$$

$$R^2 = \frac{1}{2}\varepsilon_0^2\left(\frac{R}{V}\right)^2 - \left(\frac{1}{\omega C}\right)^2.$$

Note: The graph will be R^2 versus $\left(\dfrac{R}{V}\right)^2$ and C is determined from the

Y-intercept.

(d)

No.	Resistor(s)	R (Ω)	V (V)	$\overline{P} = \dfrac{V^2}{R}$ (W)
1	R_A	680	9. 86	0. 144
2	R_B	1500	17. 36	0. 202
3	R_C	3300	22. 81	0. 159

Cont.

No.	Resistor (s)	$R\ (\Omega)$	$V\ (\mathrm{V})$	$\overline{P} = \dfrac{V^2}{R}\ (\mathrm{W})$
4	$R_A + R_B$	2180	20. 49	0. 193
5	$R_A \mathbin{/\mkern-5mu/} R_B$	468	7. 28	0. 111
6	$R_B + R_C$	4800	23. 98	0. 122
7	$R_B \mathbin{/\mkern-5mu/} R_C$	1032	13. 78	0. 186
8	$R_C + R_A$	3980	23. 66	0. 141
9	$R_C \mathbin{/\mkern-5mu/} R_A$	564	8. 42	0. 126
10	$R_A + R_B + R_C$	5480	24. 40	0. 109
11	$(R_A \mathbin{/\mkern-5mu/} R_B) + R_C$	3768	23. 43	0. 147
12	$(R_B \mathbin{/\mkern-5mu/} R_C) + R_A$	1712	18. 63	0. 202
13	$(R_C \mathbin{/\mkern-5mu/} R_A) + R_B$	2064	20. 15	0. 195
14	$(R_A \mathbin{/\mkern-5mu/} R_B) \mathbin{/\mkern-5mu/} R_C$	410	6. 22	0. 094
15	$(R_A + R_B) \mathbin{/\mkern-5mu/} R_C$	1313	16. 18	0. 200
16	$(R_B + R_C) \mathbin{/\mkern-5mu/} R_A$	596	8. 82	0. 131
17	$(R_C + R_A) \mathbin{/\mkern-5mu/} R_B$	1089	14. 36	0. 190

(data points = 17; 2. 5, >13; 2. 0, >9; 1. 5>3; 1. 0, ≤3; 0. 5)

(e)

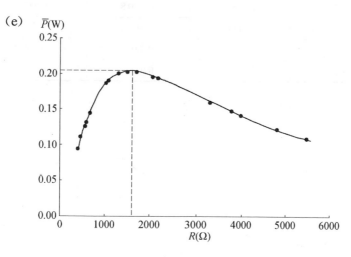

Fig. 4 – 10.

$$R \text{ at } \overline{P}_{\max} = 1600\ \Omega \Rightarrow C = \frac{1}{\omega R} = \frac{1}{2\pi \times 50 \times 1600}$$

$$= 1.9 \times 10^{-6} \text{ F} = 1.9 \ \mu\text{F}.$$

(f) Linear graph.

$R(\Omega)$	$V(\text{V})$	$\overline{P} = \dfrac{V^2}{R}(\text{W})$	$\dfrac{1}{R\overline{P}}(\Omega\text{W})^{-1}$	$\dfrac{1}{R^2}(\times 10^{-6} \ \Omega^{-2})$
410	6. 22	0. 094	0. 0259	5. 948
468	7. 28	0. 111	0. 0193	4. 565
564	8. 42	0. 126	0. 0141	3. 143
596	8. 82	0. 131	0. 0128	2. 815
680	9. 86	0. 144	0. 0102	2. 162
1032	13. 78	0. 186	0. 0052	0. 938
1089	14. 36	0. 190	0. 0048	0. 843
1313	16. 18	0. 200	0. 0038	0. 580
1500	17. 36	0. 202	0. 0033	0. 444
1712	18. 63	0. 202	0. 0029	0. 341
2064	20. 15	0. 195	0. 0025	0. 234
2180	20. 49	0. 193	0. 0024	0. 210
3300	22. 81	0. 159	0. 0019	0. 091
3768	23. 43	0. 147	0. 0018	0. 070
3980	23. 66	0. 141	0. 0018	0. 0631
4800	23. 98	0. 122	0. 0017	0. 0434
5480	24. 40	0. 109	0. 0017	0. 0333

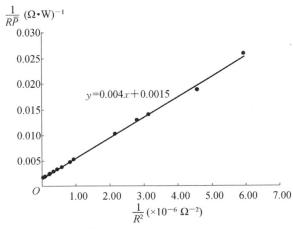

Fig. 4 – 11.

Graphical analysis: slope $= a = 0.004 \times 10^6$ Ω/W, Y-intercept $= b = 0.0015(\Omega W)^{-1}$:

$$\frac{1}{\omega^2 C^2} = \frac{a}{b} \Rightarrow C = \frac{1}{\omega}\sqrt{\frac{b}{a}} = 1.95 \times 10^{-6} \text{ F} = 1.95 \ \mu\text{F}.$$

An alternative method of linear graph

$R(\Omega)$	$V(V)$	$\bar{P} = \dfrac{V^2}{R}$ (W)	$\left(\dfrac{R}{V}\right)^2$ $(\Omega/V)^2$	$R^2 (\times 10^6 \ \Omega^2)$
410	6.22	0.094	4345	0.17
468	7.28	0.111	4133	0.22
564	8.42	0.126	4487	0.32
596	8.82	0.131	4566	0.36
680	9.86	0.144	4756	0.46
1032	13.78	0.186	5609	1.07
1089	14.36	0.190	5751	1.19
1313	16.18	0.200	6585	1.72
1500	17.36	0.202	7466	2.25
1712	18.63	0.202	8445	2.93
2064	20.15	0.195	10 492	4.26
2180	20.49	0.193	11 320	4.75
3300	22.81	0.159	20 930	10.89
3768	23.43	0.147	25 863	14.20
3980	23.66	0.141	28 297	15.84
4800	23.98	0.122	40 067	23.04
5480	24.40	0.109	50 441	30.03

Graphical analysis: Y-intercept $= \left(\dfrac{1}{\omega C}\right)^2 = 2.5428 \times 10^6 \ \Omega^2$

$$\frac{1}{\omega C} = 1.595 \times 10^3 \ \Omega \Rightarrow C = 1.99 \times 10^{-6} \text{ F} = 1.99 \ \mu\text{F}.$$

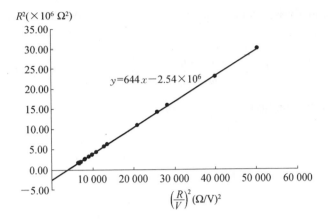

Fig. 4 - 12.

(g) Estimation of the uncertainty in the values of C obtained in (e). Estimation of the uncertainty in the values of C obtained in (f).

II. Cylindrical Bore

Background

There are many techniques to study the object with a bore inside. Mechanical oscillation method is one of the non-destructive techniques. In this problem, you are given a brass cube of uniform density with cylindrical bore inside. You are required to perform non-destructive mechanical measurements and use these data to plot the appropriate graph to find the ratio of the radius of the bore to the side of the cube.

The cube of sides a has a cylindrical bore of radius b along the axis of symmetry as shown in Fig. 4 - 13. This bore is covered by very thin discs of the same material. A, B and C represent small holes at the corners of the cube. These holes can be used for suspending the cube in two configurations. Fig. 4 - 14(a) shows the suspension using A and B; the other suspension is by using B and C as shown in Fig. 4 - 14(b).

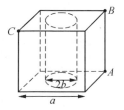

Fig. 4 – 13. Geometry of cube with cylindrical bore.

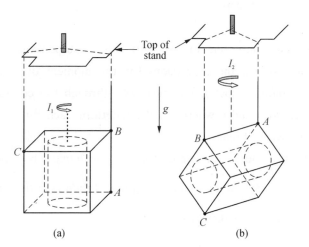

(a)   (b)

Fig. 4 – 14. Two configurations of cube's suspension.

Students may use the following in their derivation of necessary formulae:

For a solid cube of side a

$I = \dfrac{1}{6}Ma^2$ about both axes,

c. m. = centre of mass.

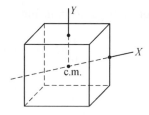

For a solid cylinder of radius b length a

$I_Y = \dfrac{1}{2}mb^2$,

$I_X = \dfrac{1}{12}ma^2 + \dfrac{1}{4}mb^2$.

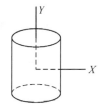

Materials and apparatus

1. Brass cube.
2. Stop watch.
3. Stand.
4. Thread.
5. Ruler/centimeter stick.
6. Linear graph papers.

Experiment

(a) Choose only one of the two bifilar suspensions as shown in Fig. 4 – 14 and derive the expressions for the moment of inertia and the period of oscillation about the vertical axis through the centre of mass in terms of l, d, b, a and g where l is the length of each thread and d is the separation between threads.

(b) Perform necessary non-destructive mechanical measurements and use these data to plot an appropriate graph and then find the value of $\dfrac{b}{a}$.

The value of g for Bangkok $= 9.78$ m/s^2.

Solution

(a) Derivation of moment of inertia I,

For configuration in Fig. 4 – 14(a),

$$I_1 = \frac{1}{6}Ma^2 - \frac{1}{2}mb^2 = \frac{1}{6}(\rho a^3)a^2 - \frac{1}{2}(\rho \pi b^2 a)b^2$$

$$= \frac{1}{6}\rho a^5 - \frac{1}{2}\rho \pi ab^4.$$

For configuration in Fig. 4 – 14(b),

$$I_2 = \frac{1}{6}Ma^2 - \frac{1}{12}ma^2 - \frac{1}{4}mb^2$$

$$= \frac{1}{6}(\rho a^3)a^2 - \frac{1}{12}(\rho \pi b^2 a)a^2 - \frac{1}{4}(\rho \pi b^2 a)b^2$$

$$= \frac{1}{6}\rho a^5 - \frac{1}{12}\rho \pi a^3 b^2 - \frac{1}{4}\rho \pi ab^4.$$

Derivation of period of oscillation T.

For both configurations:

The restoring torque $\qquad \tau = Fd$,

where $\qquad\qquad F = \dfrac{1}{2} m_0 g \dfrac{\delta s}{l}$ and $\dfrac{\delta s}{\dfrac{d}{2}} \approx \theta$,

$$F \approx \dfrac{1}{2} m_0 g \dfrac{d}{2l} \theta,$$

net mass $\qquad m_0 = \rho a^3 \left(1 - \pi \dfrac{b^2}{a^2}\right) = \rho a^3 (1 - \pi x^2)$ where $x = \dfrac{b}{a}$,

since $\qquad\qquad \tau = I\alpha, \; \alpha = \dfrac{\dfrac{1}{4} m_0 g \dfrac{d^2}{l} \theta}{I}$,

$$\omega^2 = \dfrac{4\pi^2}{T^2} = \dfrac{\dfrac{1}{4} m_0 g \dfrac{d^2}{l}}{I},$$

$$T^2 = \dfrac{4\pi^2 Il}{\dfrac{1}{4} m_0 g d^2} = \left(\dfrac{16\pi^2 I}{m_0 g d^2}\right) l.$$

For configuration in Fig. 4 – 14(a)

$$T_1^2 = \left[16\pi^2 \dfrac{\dfrac{1}{6}\rho a^5 - \dfrac{1}{2}\rho \pi a b^4}{gd^2 \quad \rho a^3 (1 - \pi x^2)}\right] l,$$

$$T_1^2 = \dfrac{8\pi^2}{3g}\left(\dfrac{a}{d}\right)^2 \dfrac{1 - 3\pi x^4}{1 - \pi x^2} l,$$

and for $d = \sqrt{2}a$, $\quad T_1^2 = \dfrac{4\pi^2}{3g} \dfrac{1 - 3\pi x^4}{1 - \pi x^2} l.$

For configuration in Fig. 4 – 14(b),

$$T_2^2 = \left[16\pi^2 \dfrac{\dfrac{1}{6}\rho a^5 \left(1 - \dfrac{\pi}{2}\dfrac{b^2}{a^2} - \dfrac{3\pi}{2}\dfrac{b^4}{a^4}\right)}{\rho a^3 (1 - \pi x^2) gd^2}\right] l,$$

$$T_2^2 = \dfrac{8\pi^2}{3g}\left(\dfrac{a}{b}\right)^2 \dfrac{1 - \dfrac{\pi x^2}{2} - \dfrac{3\pi}{2} x^4}{1 - \pi x^2} l,$$

and for $d = a$, $$T_2^2 = \frac{8\pi^2}{3g} \frac{1 - \frac{\pi x^2}{2} - \frac{3\pi}{2}x^4}{1 - \pi x^2} l.$$

(b) For configuration in Fig. 4 – 14(a), $d = 7.0$ cm,

l(cm)	T_1 for 40 oscillations(s)			T_1 (s)	$(T_1)^2$ (s^2)
16. 5	20. 60	20. 50	20. 70	0. 515	0. 265
17. 9	21. 35	21. 35	21. 30	0. 533	0. 284
22. 6	24. 05	24. 00	24. 00	0. 601	0. 362
27. 4	26. 55	26. 45	26. 55	0. 663	0. 440
29. 0	27. 40	27. 40	27. 40	0. 685	0. 469
34. 2	29. 75	29. 70	29. 65	0. 743	0. 551
36. 1	30. 60	30. 60	30. 50	0. 764	0. 584
43. 0	33. 40	33. 35	33. 50	0. 835	0. 698

3 sets of n oscillations $n \geqslant 20$,

number of lengths $l \geqslant 5$.

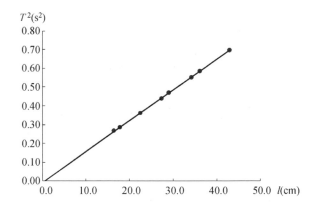

Slope of graph:

$$s_2 = \frac{0.698 - 0.265}{(43.0 - 16.5) \times 10^{-2}} = \frac{0.433}{26.5 \times 10^{-2}}$$
$$= 1.634 \text{ s}^2/\text{m},$$

$$x = \frac{b}{a} = 0.25.$$

For configuration in Fig. $4 - 14(b)$, $d = 4.9$ cm.

l(cm)	T_1 for 50 oscillations(s)			T_1 (s)	$(T_1)^2$ (s^2)
43.8	46.95	46.90	46.80	0.938	0.880
36.0	42.70	42.45	42.50	0.851	0.724
30.9	39.60	39.40	39.35	0.789	0.623
26.5	36.40	36.30	36.45	0.728	0.530
19.5	30.80	30.85	30.75	0.616	0.379

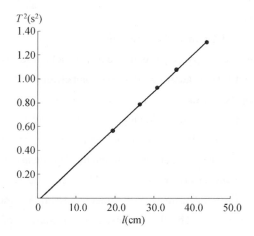

Slope of graph:

$$s_2 = \frac{0.88 - 0.53}{(43.8 - 26.5) \times 10^{-2}} = \frac{0.35}{17.3 \times 10^{-2}} = 2.02 \text{ s}^2/\text{m},$$

$$x = \frac{b}{a} = 0.22.$$

Minutes of the Fifth Asian Physics Olympiad

Hanoi (Vietnam) April 26 – May 4, 2004

Present: Prof. Tran Van Nhung (Chairman)

Prof. Waldemar Gorzkowski (President of the IPhO)

Prof. Yohanes Surya (President of the APhO)

Prof. Ming-Juey Lin (Secretary of the APhO)

Prof. Phan Hong Khoi (Honourable President of the APhO5)

Prof. Dam Trung Don (President of the APhO5)

Mr. Phan Duy Nga (Secretary)

Leaders

Observers

Members of Organizing Committee

1. The total number of participants in the 5th APhO is 148. Among them there are 97 official competitors, 6 guest (unofficial) competitors, 9 female competitors, 27 leaders (including two leaders of the guest team), 16 observers and 2 visitors. These participants come from the following 13 countries and territories: Australia, Cambodia, Chinese Taipei, Indonesia, Israel, Kazakhstan, Malaysia, Mongolia, People's Republic of China, Singapore, Thailand, Turkmenistan, and Vietnam. Cambodia and Turkmenistan are participating APhO for the first time. Indonesia sends two teams: one official team with 8 students (Team A) and one guest team with 6 students (Team B). The Turkmenistan team consists of 4 students, the Kazakhstan team — 5 students. The youngest participant comes from Indonesia.

2. The opening ceremony of the 5th APhO was honored by presence of H. E. Mr. Tran Duc Luong, President of the Socialist Republic of Vietnam. The Vice Premier of the Socialist Republic of Vietnam, Dr. Pham Gia Khiem will be the Guest of Honour for the closing ceremony.

3. The results of the marking papers are as follow:

The best score (45.10 points) was achieved by RUITIAN LANG from People's Republic of China.

The second score (44.80 points) by ZHEN LI from People's Republic of China.

The third score (42.40 points) by YALI MIAO from People's Republic of

China.

According to the statutes of APhO, the following limits for awarding medals and honorable mentions were established:

Gold medal:	39 points,
Silver medal:	34 points,
Bronze medal:	28 points,
Honorable mention:	22 points.

According to the above limits, 6 Gold medals, 7 Silver medals, 9 Bronze medals, and 23 Honorable mentions were awarded for the competitors of the 5th APhO (see an attachment list).

4. In addition to the regular prizes, a number of special prizes are awarded:

● for the Absolute Winner (45.10 points): Ruitian Lang from People's Republic of China

● for the best score in theory (27. 90 points): Ruitian Lang from People's Republic of China

● for the best score in experiment (19. 75 points): Yeming Shi from People's Republic of China.

A number of special mentions will be awarded:

● for the best female participant: Veronika Ulychny from Israel

● for the youngest participant: Yongki Utama from Indonesia

● for the best participant from the new participating countries:

(1) Ear Socheat from Cambodia

(2) Rustam Durdyyew Halmuhammedowiç. from Turkmenistan.

5. The International Board agreed to publish the best or the most creative solutions in the APhO proceeding without asking the student's permission.

6. Chairman Tran Van Nhung informed that the equipment used for the experiment exam would be distributed to the teams as souvenirs.

7. A list of the Hosts of the APhO in the future is as follows:

● 2005 – Indonesia

● 2006 – Kazakhstan

● 2007 – People's Republic of Chine (tentative)

● 2008 – Australia (tentative)

● 2009 – Uzbekistan

● 2010 – Malaysia (tentative)

- 2011 – Israel.

8. Acting on behalf of the organizers of the 6th APhO, the Indonesian delegation announced that the 6th APhO would be organized in Pekanbaru (Riau province Indonesia) in April 2005 and cordially invited all participating countries to attend the competition.

9. Prof. Yohanes Surya (President of the APhO) and Prof. Waldemar Gorzkowski (Honourable President of the APhO), acting on behalf of all the members of the International Board, expressed deep thanks to Prof. Tran Van Nhung, Deputy Minister of Education and Training of Vietnam, Chairman, and members of the 5th APhO Organizing Committee, for excellent organization and execution of the 5th APhO.

Hanoi, May 2, 2004

Mr. Phan Duy Nga,	**Prof. Ming-Juey Lin**	**Prof. Yohanes Surya,**
Executive Secretary of	Secretary of the APhO	President of the APhO
the 5th APhO		

Theoretical Competition

April 28, 2004

Problem 1
Measure the Mass in the Weightless State

In the spacecraft orbiting the Earth, there is weightless state, so that one cannot use ordinary instrument to measure the weight and then to deduce the mass of astronaut. Skylab 2 and some other spacecrafts are supplied with a Body Mass Measurement Device, which consists of a chair attached to one end of a spring. The other end of the spring is attached to a fixed point of the spacecraft. The axis of the spring passes through the center of mass of the craft. The force constant (the hardness) of the spring is $k = 605.6$ N/m.

(1) When the craft is fixed on the pad, the chair (without person) oscillates with the period $T_0 = 1.281\ 95$ s.

Calculate the mass m_0 of the chair.

(2) When the craft orbits the Earth the astronaut straps himself into the chair and measures the period T' of the chair oscillation. He obtains $T' = 2.330\ 44$ s, then calculates roughly his mass. He feels some doubt and tries to find the true value of his mass. He measures again the period of oscillation of the chair (without person), and find $T'_0 = 1.273\ 95$ s.

What is the true value of the astronaut's mass and the craft's mass?

Note: Mass of spring is negligible and the astronaut is floating.

Solution

(1) Formula for period of oscillation:

$$T_0 = \frac{2\pi}{\omega_0} = 2\pi\sqrt{\frac{m_0}{k}}. \tag{1}$$

One can deduce

$$m_0 = \frac{k}{4\pi^2} T_0^2 = 2521 \text{ kg.} \tag{2}$$

(2) When the craft orbits the Earth, oscillating system is the spring with one end attached to the chair of mass m_0 and the other end attached to the craft with mass M. This system oscillates like an object with mass

$$m'_0 = \frac{m_0 M}{m_0 + M} \tag{3}$$

attached to an end of the spring, the other end of the spring is fixed (m'_0 is the reduced mass of the system craft-chair). The period T'_0 of the system is also given by (1) and (2). We can deduce

$$\frac{m_0}{m'_0} = \left(\frac{T_0}{T'_0}\right)^2 = \left(\frac{1.281\ 95}{1.273\ 95}\right)^2. \tag{4}$$

The mass M of the craft can be calculated from (3)

$$M = \frac{m_0}{\frac{m_0}{m'_0} - 1} = \frac{25.21}{\left(\frac{T_0}{T'_0}\right)^2 - 1}$$

$$= \frac{25.21}{\left(\frac{1.281\ 95}{1.273\ 95}\right)^2 - 1} = 2001 \text{ kg.} \tag{5}$$

Let m be the mass of astronaut and chair, the coressponding reduce mass is m':

$$m' = \frac{mM}{m + M}, \tag{6}$$

the expression of m is then

$$m = \frac{m'}{1 - \frac{m'}{M}}.$$

The reduce mass m' can be calculated from the oscillation period T' by using formula (2):

$$m' = \frac{605.6}{4} \cdot \left(\frac{2.330\ 44}{3.141\ 6}\right)^2 = 83.31 \text{ kg,}$$

the true value of the mass m is

$$m = \frac{83.31}{1 - \dfrac{83.31}{2001}} = 86.93 \text{ kg},$$

the true value of the mass of astronaut:

$$86.93 - 25.21 = 61.72 \text{ kg}.$$

Problem 2
Optical Fiber

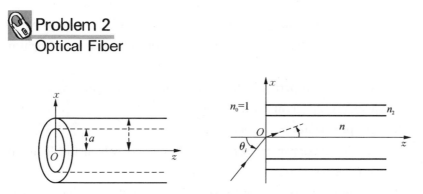

Fig. 5 - 1.

An optical fiber consists of a cylindrical core of radius a, made of a transparent material with refraction index varying gradually from the value $n = n_1$ on the axis to $n = n_2$ (with $1 < n_2 < n_1$) at a distance a from the axis, according to the formula

$$n = n(x) = n_1 \sqrt{1 - \alpha^2 x^2}$$

where x is the distance from the core axis and α is a constant. The core is surrounded by a cladding made of a material with constant refraction index n_2. Outside the fiber is air, of refractive index n_0.

Let Oz be the axis of the fiber, with O the center of the fiber end.

Give $n_0 = 1.000$; $n_1 = 1.500$; $n_2 = 1.460$, $a = 25$ μm.

(1) A monochromatic light ray enters the fiber at point O under an incident angle θ_i, the incident plane being the plane xOz.

(a) Show that at each point on the trajectory of the light in the fiber, the refractive index n and the angle θ between the light ray and the Oz axis satisfy the relationship $n\cos\theta = C$ where C is a constant. Find the expression for C in terms of n_1 and θ_i.

(b) Use the result found in (1) (a) and the trigonometric relation $\cos\theta = (1+\tan^2\theta)^{-\frac{1}{2}}$, where $\tan\theta = \dfrac{\mathrm{d}x}{\mathrm{d}z} = x'$ is the slope of the tangent to the trajectory at point (x, z), derive an equation for x'. Find the full expression for α in terms of n_1, n_2 and a. By differentiating the two sides of this equation versus z, find the equation for the second derivative x''.

(c) Find the expression of x as a function of z, that is $x = f(z)$, which satisfies the above equation. This is the equation of the trajectory of light in the fiber.

(d) Sketch one full period of the trajectories of the light rays entering the fiber under two different incident angles θ_i.

(2) Light propagates in the optical fiber.

(a) Find the maximum incident angle $\theta_{i\,m}$, under which the light ray still can propagate inside the core of the fiber.

(b) Determine the expression for coordinate z of the crossing points of a light ray with Oz axis for $\theta_i \neq 0$.

(3) The light is used to transmit signals in the form of very short light pulses (of negligible pulse width).

(a) Determine the time τ it takes the light to travel from point O to the first crossing point with Oz for incident angle $\theta_i \neq 0$ and $\theta_i \leqslant \theta_{im}$.

The ratio of the coordinate z of the first crossing point and τ is called the propagation speed of the light signal along the fiber. Assume that this speed varies monotonously with θ_i.

Find this speed (called v_m) for $\theta_i = \theta_{im}$.

Find also the propagation speed (called v_0) of the light along the axis Oz.

Compare the two speeds.

(b) The light beam bearing the signals is a converging beam entering the fiber at O under different incident angles θ_i with $0 \leqslant \theta_i \leqslant \theta_{iM}$. Calculate the highest repetition frequency f of the signal pulses, so that at a distance $z = 1000$ m two consecutive pulses are still separated (that is, the pulses do not overlap).

Attention

1. The wave properties of the light are not considered in this problem.
2. Neglect any chromatic dispersion in the fiber.
3. The speed of light in vacuum is $c = 2.998 \times 10^8$ m/s.
4. You may use the following formulae:

The length of a small arc element ds in the xOz plane is

$$ds = dz \sqrt{1 + \left(\frac{dx}{dz} \right)^2} \; ;$$

$$\int \frac{dx}{\sqrt{a^2 - b^2 x^2}} = \frac{1}{b} \text{Arcsin} \frac{bx}{a} \; ;$$

$$\int \frac{x^2 \, dx}{\sqrt{a^2 - b^2 x^2}} = -\frac{x \sqrt{a^2 - b^2 x^2}}{2b^2} + \frac{a^2 \text{Arcsin} \frac{bx}{a}}{2b^3} \; ;$$

Arcsin x is the inverse function of the sine function. Its value equals the less angle the sine of which is x. In other words, if $y = \arcsin x$, $\sin y = x$.

Solution

(1) (a) At both sides of the point O (outside and inside the fiber), according to Snell law, we have

$$n_0 \sin \theta_i = n_1 \sin \theta_1 , \tag{1}$$

where θ_1 is the value of angle θ at point O inside the fiber.

The light trajectory lays in the xOz plane. Because the refraction index n varies along x direction, we divide Ox axis into small elements dx, so that in each of these elements n can be considered as constant. We have, then

$$n \sin i = (n + dn) \cdot \sin(i + di), \tag{2}$$

where i is the angle between the light trajectory and x direction. Because $\theta + i = \frac{\pi}{2}$,

$$n \cos \theta = (n + dn) \cdot \cos(\theta + d\theta). \tag{3}$$

Thus, at each point of coordinate x on the light trajectory, we have

$$n\cos\theta = n_1\sqrt{1-\alpha^2 x^2}\cos\theta = n_1\cos\theta_1. \tag{4}$$

Because

$$\cos\theta_1 = \sqrt{1-\sin^2\theta_1} = \sqrt{1-\frac{\sin^2\theta_i}{n_1^2}}, \tag{5}$$

we have

$$n\cos\theta = n_1\cos\theta_1 = n_1\sqrt{1-\frac{\sin^2\theta_i}{n_1^2}} = \sqrt{n_1^2-\sin^2\theta_i}.$$

Then

$$n\cos\theta = C = \sqrt{n_1^2-\sin^2\theta_i}. \tag{6}$$

(1) (b) Because $\dfrac{dx}{dz} = x' = \tan\theta$, from (6) we have:

$$n_1\sqrt{1-\alpha^2 x^2}\cos\theta = n_1\sqrt{1-\alpha^2 x^2}(1+\tan^2\theta)^{-\frac{1}{2}} = C. \tag{7}$$

Squaring the two sides, we obtain

$$(1-\alpha^2 x^2)(1+\tan^2\theta)^{-1} = \frac{C^2}{n_1^2}$$

and

$$1+x'^2 = (1-\alpha^2 x^2)\frac{n_1^2}{C^2}. \tag{8}$$

After derivating the two sides of (8) versus z, we get

$$x'' + \frac{\alpha^2 n_1^2}{C^2}x = 0. \tag{9}$$

Because $n = n_1\sqrt{1-\alpha^2 x^2}$ and

$n = n_1$ at $x = 0$,

$n = n_2$, at $x = a$,

we get

$$\alpha = \frac{\sqrt{n_1^2-n_2^2}}{a\cdot n_1}.$$

Finally, we get the equation for x'':

$$x'' + \frac{n_1^2 - n_2^2}{a^2(n_1^2 - \sin^2\theta_i)} \cdot x = 0. \tag{10}$$

(c) The equation for the light trajectory is obtained by solving (10). This is an equation similar to that for an harmonic oscillation, which solution can be written right away

$$x = x_0 \sin(pz + q) \tag{11}$$

with

$$p = \frac{1}{a}\sqrt{\frac{n_1^2 - n_2^2}{n_1^2 - \sin^2\theta_i}}.$$

The parameters p and q are determined from the boundary conditions:

at $z = 0$, $x = 0$, hence $q = 0$;

at $z = 0$ inside the fiber, $x' = \dfrac{dx}{dz} = \tan\theta_1$, then

$$x_0 = \frac{\tan\theta_1}{p}$$

$$= \frac{a \cdot \sin\theta_i}{\sqrt{n_1^2 - n_2^2}}. \tag{12}$$

The equation for the trajectory of the light inside the fiber is

$$x = \frac{a\sin\theta_i}{\sqrt{n_1^2 - n_2^2}} \cdot \sin\left(\sqrt{\frac{n_1^2 - n_2^2}{n_1^2 - \sin^2\theta_i}} \cdot \frac{z}{a}\right). \tag{13}$$

(d) Here is a sketch of the trajectories of two rays entering the fiber at O, under different incident angles.

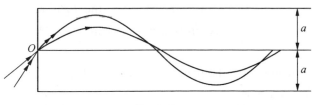

Fig. 5 – 2.

(2) (a) The condition for the light to propagate along the fiber is that

$x_0 \leqslant a$. This means

$$\frac{a\sin\theta_i}{\sqrt{n_1^2 - n_2^2}} \leqslant a$$

or

$$\sin\theta_i \leqslant \sqrt{n_1^2 - n_2^2}. \qquad (14)$$

Thus the incident angle θ_i must not exceed θ_{im}, with

$$\sin\theta_{im} = \sqrt{n_1^2 - n_2^2} = 0.344 \qquad (14a)$$

or

$$\theta_i \leqslant \theta_{im} = \arcsin(\sqrt{n_1^2 - n_2^2}) = \arcsin 0.344$$
$$= 0.351\,\text{rad} = 20.13°.$$

(b) The crossing points of the light beam with Oz axis must satisfy the condition $pz = k\pi$, with k — an integer. The z coordinates of these points are

$$z = \frac{k\pi}{p} = k\pi a \sqrt{\frac{n_1^2 - \sin^2\theta_i}{n_1^2 - n_2^2}} \qquad (15)$$

except for $\theta_i = 0$.

(3) (a) The rays entering the fiber at different incident angles have different trajectories. As a consequence, the propagation speeds of the rays along the fiber should be different.

The light trajectories are sinusoidal as given in (13). Let us calculate the time τ it takes the light to propagate from point O to its first crossing point with Oz axis. This is twice the time it takes the light to propagate from point O to its position most distant from Oz axis.

The time required for the light to travel a small segment ds along its trajectory is

$$dt = \frac{n}{c}ds = \frac{n}{c}\sqrt{dx^2 + dz^2} = \frac{n}{c}\sqrt{1 + \frac{dz^2}{dx^2}} \cdot dx$$

$$= \frac{n}{c} \sqrt{1 + \left(\frac{1}{\tan \theta}\right)^2} \cdot dx = \frac{n}{c} \cdot \frac{dx}{\sin \theta}.$$

From (6), we have

$$dt = \frac{n_1^2 (1 - \alpha^2 x^2)}{c \cdot \sqrt{\sin^2 \theta_i - n_1^2 \alpha^2 x^2}} \cdot dx$$

and

$$\frac{\tau}{2} \int_0^{x_0} dt = \frac{n_1^2}{c} \left(\int_0^{x_0} \frac{dx}{\sqrt{\sin^2 \theta_i - n_1^2 \alpha^2 x^2}} - \alpha^2 \int_0^{x_0} \frac{x^2 \, dx}{\sqrt{\sin^2 \theta_i - n_1^2 \alpha^2 x^2}} \right)$$

$$= \frac{n_1^2}{c} (I_1 - \alpha^2 I_2), \tag{16}$$

where

$$I_1 = \frac{1}{n_1 \alpha} \text{Arcsin} \frac{n_1 \alpha x}{\sin \theta_i} \Big|_0^{x_0} = \frac{\pi \alpha}{2 \sqrt{n_1^2 - n_2^2}}, \tag{17}$$

$$I_2 = \frac{-x \sqrt{\sin^2 \theta_i - n_1^2 \alpha^2 x^2}}{2 n_1^2 \alpha^2} \Big|_0^{x_0} + \frac{\sin^2 \theta_i \cdot \text{Arcsin} \frac{n_1 \alpha x}{\sin \theta_i}}{2 n_1^3 \alpha^3} \Big|_0^{x_0}$$

$$= \frac{\pi \sin^2 \theta_i}{4 n_1^3 \alpha^3}. \tag{18}$$

Using (16), (17), (18), we obtain

$$\tau = \frac{\pi \alpha \cdot n_1^2}{c \sqrt{n_1^2 - n_2^2}} \left(1 - \frac{\sin^2 \theta_i}{2 n_1^2}\right). \tag{19}$$

The propagation speed along the fiber is $v = \frac{z}{\tau}$, where z is the coordinate of the first crossing point, which is determined by (15) for $k = 1$. Because z and τ depend on the incident angle θ_i, v also depends on θ_i.

For $\theta_i = \theta_{im}$, from (14a), we get

$$v_m = \frac{\pi \alpha n_2}{\sqrt{n_1^2 - n_2^2}} \cdot \frac{2c \sqrt{n_1^2 - n_2^2}}{\pi \alpha n_1^2} \left(1 + \frac{n_2^2}{n_1^2}\right)^{-1} = \frac{2c n_2}{n_1^2 + n_2^2} \tag{20}$$

and

$$v_m = \frac{2 \times 2.998 \times 10^8 \times 1.460}{1.500^2 + 1.460^2} = 1.998 \times 10^8 \text{ m/s}. \tag{20a}$$

The propagation speed of the light along the Oz axis is

$$v = \frac{c}{n_1} \tag{21}$$

because the refraction index is n_1 on the axis of the fiber.

The numerical value is

$$v_0 = \frac{2.998 \times 10^8}{1.5} = 1.999 \times 10^8 \text{ m/s.} \tag{21a}$$

(3) (b) If the beam of the light pulses is formed by rays converging at O, then the rays with different incident angles has different propagation speeds. The two rays of incident angles $\theta_i = 0$ and $\theta_i = \theta_{iM}$ arrive to the plane z with a time delay

$$\Delta t = \frac{z}{v_m} - \frac{z}{v_0} = \frac{z}{c} \cdot \frac{(n_1 - n_2)^2}{2n_2}. \tag{22}$$

This means that a very short light pulse becomes a pulse of finite width Δt given by (22) at the plane z. If two consecutive pulses enter the fiber with a delay greater than Δt, then at the plane z, they are separated. Hence the repetition frequency of the pulses must not exceed the maximal value

$$f_m = (\Delta t)^{-1} = \frac{2 \cdot c \cdot n_2}{z \cdot (n_1 - n_2)^2}. \tag{23}$$

If $z = 1000$ m , then

$$f_m = \frac{2 \times 2.998 \times 10^8 \times 1.460}{1000 \times (1.500 - 1.460)^2} = 547.1 \text{ MHz.}$$

Problem 3
Compression and Expansion of a Two Gases System

A cylinder is divided in two compartments with a mobile partition NM. The compartment in the left is limited by the fond of the cylinder and the partition NM (Fig. 5 – 3). This compartment contains one mole of water vapor.

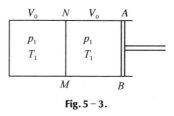

Fig. 5 – 3.

The compartment in the right is limited by the partition NM and a mobile piston AB. This compartment contains one mole of nitrogen gas (N_2).

At first, the volumes and temperatures of the gases in two compartments are equal. The partition NM is well heat conductive. Its heat capacity is very small and can be neglected.

The specific volume of liquid water is negligible in comparison with the specific volume of water vapor at the same temperature.

The specific latent heat of vaporization L is defined as the amount of heat that must be delivered to one unit of mass of substance to convert it from liquid state to vapor at the same temperature. For water at $T_0 = 373$ K, $L = 2250$ kJ/kg.

(1) Suppose that the piston and the wall of the cylinder are heat conductive, the partition NM can slide freely without friction. The initial state of the gases in the cylinder is defined as follows:

Pressure $p_1 = 0.5$ atm. ; total volume $V_1 = 2V_0$; temperature $T_1 = 373$ K.

The piston AB slowly compresses the gases in a quasi-static (quasi-equilibrium) and isothermal process to the final volume $V_F = V_0/4$.

(a) Draw the $p(V)$ curve, which is the curve representing the dependence of pressure p on the total volume V of both gases in the cylinder at temperature T_1. Calculate the coordinates of important points of the curve.

Gas constant: $R = 8.31$ J/(mol $\cdot$ K) or $R = 0.0820$ (L $\cdot$ atm.)/(mol $\cdot$ K); 1 atm. $= 101.3$ kPa;

Under the pressure $p_0 = 1$ atm. , water boils at the temperature $T_0 = 373$ K.

(b) Calculate the work done by the piston in the process of gases compressing.

$$\int \frac{dV}{V} = \ln V.$$

(c) Calculate the heat delivered to outside in the process.

(2) All conditions as in (1) except that there is friction between

partition NM and the wall of the cylinder so that NM displaces only when the difference of the pressures acting on its two opposed faces attains 0.5 atm. and over (assuming that the coefficients of static and kinetic friction are equal).

(a) Draw the $p(V)$ curve representing the pressure p in the right compartment as a function of the total volume V of both gases in the cylinder at a constant temperature T_1.

(b) Calculate the work done by the piston in compressing the gases.

(c) After the volume of gases reaches the value $V_F = V_0/4$, piston AB displaces slowly to the right and makes a quasi-static and isothermal process of expansion of both substances (water and nitrogen) to the initial total volume $2V_0$. Continue to draw in the diagram in question (2) (a) the curve representing this process.

Hint for (2)

Create a table like the one shown here and use it to draw the curves as required in (2) (a) and (2) (c).

State	Left compartment		Right compartment		Total volume	Pressure on piston AB
	Volume	Pressure	Volume	Pressure		
initial	V_0	0.5 atm.	V_0	0.5 atm.	$2V_0$	0.5 atm.
2						
3						
·						
·						
·						
·						
·						
final					$2V_0$	

(3) Suppose that the wall and the fond of the cylinder and the piston are heat insulator, the partition NM is fixed and heat conductive, the initial state of gases is as in (1). Piston AB moves slowly toward the right side and increases the volume of the right compartment until the water vapor begins

to condense in the left compartment.

(a) Calculate the final volume of the right compartment.

(b) Calculate the work done by the gas in this expansion.

The ratio of isobaric heat capacity to isochoric one $\gamma = \dfrac{C_p}{C_V}$ for nitrogen

is $\gamma_1 = \dfrac{7}{5}$ and for water vapor $\gamma_2 = \dfrac{8}{6}$.

In the interval of temperature from 353 K to 393 K one can use the following approximate formula:

$$p = p_0 \exp\left[-\frac{\mu L}{R}\left(\frac{1}{T} - \frac{1}{T_0}\right)\right]$$

where T is boiling temperature of water under pressure p, μ its molar mass. p_0, L_0 and T_0 are given above.

Solution

(1) (a) The isotherm curve is shown in Fig. 5 - 4.

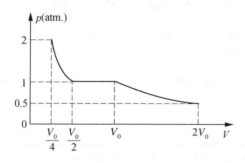

Fig. 5 - 4.

$$V_0 = \frac{RT_1}{p_1} = \frac{8.31 \times 373}{0.5 \times 1.013 \times 10^5} = 0.0612 \text{ m}^3 = 61.2 \text{ dm}^3.$$

(b) The process of compressing can be divided into 3 stages:

$$(p_1, 2V_0) \rightarrow (2p_1, V_0) \rightarrow \left(2p_1, \frac{V_0}{2}\right) \rightarrow \left(4p_1, \frac{V_0}{4}\right).$$

$$(1) \qquad\qquad (2) \qquad\qquad (3) \qquad\qquad (4)$$

The work in each stage can be calculated as follows:

$$A_{12} = -\int_{2V_0}^{V_0} p \, dV = 2RT_1 \int_{V_0}^{2V_0} \frac{dV}{V} = 2RT_1 \ln 2 = 4297 \text{ J},$$

$$A_{23} = 2p_1 \left(V_0 - \frac{V_0}{2} \right) = RT_1 = 3100 \text{ J},$$

$$A_{34} = -\int_{\frac{1}{2}V_0}^{\frac{1}{4}V_0} p' \, dV = RT_1 \int_{\frac{1}{4}V_0}^{\frac{1}{2}V_0} \frac{dV}{V} = RT_1 \ln 2 = 2149 \text{ J}.$$

The total work of gases compressing is

$$A = A_{12} + A_{23} + A_{34} = 9545 \text{ J} \simeq 9.55 \text{ kJ}. \tag{1}$$

(c) In the second stage 23, all the water vapor (one mole) condenses. The heat Q' delivered in the process equals the sum of the work A and the decrease ΔU of internal energy of one mole of water vapor in the condensing process.

$$Q' = \Delta U + A_{12} + A_{23} + A_{34}.$$

One can remark that $\Delta U + A_{23}$ is the heat delivered when one mole of water vapor condenses, and equals $0.018 \times L$. Thus

$$Q' = \Delta U + A_{11} + A_{23} + A_{34} = 0.018 \times L + A_{11} + A_{34}$$
$$= 46.946 \text{ J} \simeq 47 \text{ kJ}. \tag{2}$$

(2) The process of compression (2) (a) and expansion (2) (c) of gases can be divided into several stages. The stages are limited by the following states:

State	Left compartment		Right compartment		Total volume	Pressure on piston (atm.)
	Volume	Pressure (atm.)	Volume	Pressure (atm.)		
1	V_0	0.5	V_0	0.5	$2V_0$	0.5
2	V_0	0.5	$0.5V_0$	1	$1.5V_0$	1
3	$0.5V_0$	1	$\frac{V_0}{3}$	1.5	$\frac{5}{6}V_0$	1.5
4	0	1	$\frac{V_0}{3}$	1.5	$\frac{V_0}{3}$	1.5

Cont.

State	Left compartment		Right compartment		Total volume	Pressure on piston (atm.)
	Volume	Pressure (atm.)	Volume	Pressure (atm.)		
5	0	1.5	$\dfrac{V_0}{4}$	2	$\dfrac{V_0}{4}$	2
6	0	1.5	$\dfrac{V_0}{3}$	1.5	$\dfrac{V_0}{3}$	1.5
7	0	1	V_0	0.5	V_0	0.5
8	$0.5V_0$	1	V_0	0.5	$1.5V_0$	0.5
9	$(2-\sqrt{2})V_0$	$\dfrac{\sqrt{2}+2}{4}$	$\sqrt{2}V_0$	$\dfrac{\sqrt{2}}{4}$	$2V_0$	$\dfrac{\sqrt{2}}{4} \approx 0.35$

(a) See Fig. 5 – 5 below.

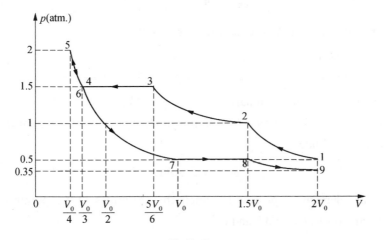

Fig. 5 – 5.

(b) The work A_p done by the piston in the process of compressing the gases equals the sum of the work A calculated in (1) and the work done by the force of friction. The latter equals 0.5 atm. $\times V_0 = p_1 V_0$ (the force of kinetic friction appears in the process 234 during which the displacement of the partition NM corresponding to a variation V_0 of volume of the left compartment). Then, we have:

$$A_p = A + p_1 V_0 = 9545 + 8.31 \times 373 = 12\,645 \text{ J} \cong 12.65 \text{ kJ}.$$

(c) In process 89 the pressure in the left compartment is always larger than the pressure in the right one (with a difference of 0.5 atm.). If p denotes the pressure in the right compartment, the pressure in the left one will be $p + 0.5$ atm.

Let V be the total volume in process 89, we have

$$\frac{RT_1}{p} + \frac{RT_1}{p + 0.5} = V$$

with $V = 2V_0 = \dfrac{2RT_1}{0.5}$, p can be defined by $\dfrac{RT_1}{p} + \dfrac{RT_1}{p + 0.5} = \dfrac{2RT_1}{0.5}$.

This is equivalent to:

$$p + 0.5 + p = 4p \cdot (p + 0.5),$$

$$p = \frac{1}{\sqrt{8}} = \frac{\sqrt{2}}{4} = 0.35 \text{ atm.}.$$

The pressure in the right compartment is $p = 0.35$ atm..

The volume of the right compartment is $\sqrt{2}V_0$.

The pressure in the left compartment is $p + 0.35 = 0.85$ atm..

The volume of the left compartment is $(2 - \sqrt{2})V_0$.

(3) (a) Apply the first law of thermodynamic for the system of two gases in the cylinder:

$$\delta q = dU + \delta A. \tag{3}$$

In an element of process in which the variations of temperature and volume are respectively dT and dV:

$$\delta q = 0; \quad dU = (C_{V_1} + C_{V_2})dT = \left(\frac{R}{\gamma_1 - 1} + \frac{R}{\gamma_2 - 1}\right)dT;$$

$$\delta A = p \cdot dV.$$

On the other hand, $\qquad pV = RT. \tag{4}$

One can deduce the differential equation for the process:

$$\left(\frac{R}{\gamma_1 - 1} + \frac{R}{\gamma_2 - 1}\right)dT + p \cdot dV = 0$$

or

$$\frac{dT}{T} + \frac{(\gamma_1 - 1)(\gamma_2 - 2)}{\gamma_1 + \gamma_2 - 2} \frac{dV}{V} = 0. \tag{5}$$

By putting

$$K = \frac{(\gamma_1 - 1)(\gamma_2 - 1)}{\gamma_1 + \gamma_2 - 2} = \frac{2}{11} \tag{6}$$

after integrating (5), we have

$$TV^K = \text{const.} \tag{7}$$

The condensing temperature of water vapor under the pressure 0.5 atm. is also the boiling temperature T' of water under the same pressure. By using the given approximate formula, we obtain

$$\frac{1}{T'} - \frac{1}{T_0} = \left(-\frac{R}{\mu L}\right) \ln \frac{p}{p_0}.$$

- If we consider p approximatively constant (with relative deviation about $\frac{20}{373} \approx 5\%$) T' can be easily found, and

$$T' = 354 \text{ K}.$$

The volume V' of the right compartment at temperature T' can be calculated as follows:

$$V' = V_0 \left(\frac{T_1}{T'}\right)^{\frac{1}{K}} = V_0 \left(\frac{373}{354}\right)^{\frac{11}{2}} = 1.33 V_0 \cong 1.3 V_0$$

$$= 81.6 \text{ dm}^3 = 0.0816 \text{ m}^3 \cong 0.08 \text{ m}^3.$$

(b) The work done by the gas in the expansion is

$$A = -\Delta U = (C_{V_1} + C_{V_2})(T_0 - T')$$

$$= \left(\frac{R}{\gamma_1 - 1} + \frac{R}{\gamma_2 - 2}\right)(373 - 354)$$

$$= \left(\frac{5}{2}R + \frac{6}{2}R\right) \times 19 = 868 \text{ J} \cong 9 \times 10^2 \text{ J}.$$

- If we consider the dependence of water vapor pressure p on temperature T', we must resolve the transcendental equation

$$\frac{1}{T'} - \frac{1}{T_0} = \left(-\frac{R}{\mu L}\right) \ln \frac{1}{2} \frac{T'}{T_0} = \frac{R}{\mu L} \ln 2 - \frac{R}{\mu L} \ln \frac{T'}{T_0}.$$

This equation can be reduced to a numerical one:

$$\frac{1}{T'} - \frac{1}{373} = 1.422 \times 10^{-4} - 2.052 \times 10^{-4} \ln \frac{T'}{373}.$$

By giving T' different values 354, 353, 352 and choosing the one which satisfies this equation, we find the approximate solution:

$$T' = 353 \text{ K.}$$

With this value of temperature, the volume of the right compartment is

$$V' = V_0 \left(\frac{373}{353}\right)^{\frac{11}{2}} = 1.35 V_0 = 0.082 \text{ m}^2 \cong 0.08 \text{ m}^3.$$

(c) The work done by the gas in the expansion is

$$A = \frac{11}{2} R \times 20 = 914 \text{ J} \cong 9 \times 10^2 \text{ J.}$$

Experimental Competition

April 30, 2004

Problem 1
Hall Effect and Magnetoresistivity Effect

Apparatus and materials

(1) Three digital multimeters.

(2) A Hall sensor with four pins $MNPQ$ (M in black wire, N yellow wire, P red wire, Q green wire), fixed on a printed circuit, a pair of conductors leading to M, N; another pair of conductors leading to P, Q.

(3) A permanent magnet in the shape of a disk, of radius $r = 14$ mm, of thickness $t = 4$ mm. The magnetization is perpendicular to the surface of the disk. The value B_0 (in Tesla) of the magnetic field at the surface of the magnet is written on its surface.

During the experiment, keep the magnet far away from the Hall sensor whenever you do not use it.

(4) A coil of N turns is wound on a core having the shape of a toroid, made of a ferromagnetic material. The average radius of the core is $\rho = 25$ mm. The toroid has a gap of width $d = 3$ mm.

(5) A box with two independent 1.5 V dry cells. The cell connected in series to a 10 kΩ variable resistor, called battery 1, is used to supply the current to the Hall sensor. The second cell, called battery 2, is used to supply the current to the coil only during the measurement.

(6) A protractor with a small hole at its center.

(7) A piece of plexiglass with a small needle fixed on it.

(8) A holder for the printed circuit with the Hall sensor.

(9) A small piece of plastic used to fix the sensor on the needle.

(10) Conductors with negligible resistance.

(11) Graph papers.

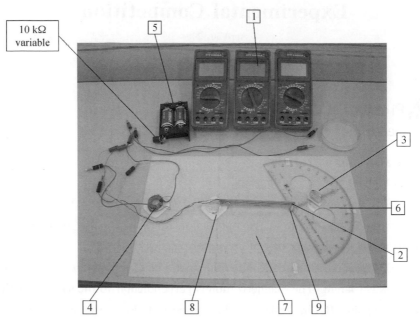

10 kΩ variable

Fig. 5 - 6.

Experiment

Ⅰ. Introduction

The magnetoresistivity effect and the Hall effect.

Consider a conductor sample in the shape of a parallelopiped of length a, width b and thickness c (see Fig. 5 - 7). The current I flows along the direction of a. If the sample is placed in a magnetic field $\boldsymbol{B}$, The magnetic field affects the resistance R of the sample. This effect is called magneto - resistivity effect (MRE). Let ΔR the increase of the resistance R of the sample, R_0— the value of R in the absence of magnetic field, then the magnitude of the MRE is defined by the ratio $\dfrac{\Delta R}{R_0}$.

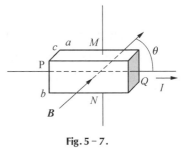

Fig. 5 - 7.

Assume that the applied magnetic field is uniform and the magnetic induction vector B is parallel to the top face of the sample as shown in Fig.5 – 7. If the charge carriers in the sample are electrons, the Lorentz force will bend them upward, and the top face of the sample will be charged negatively. This effect is called the Hall effect. The voltage appearing between electrodes M (on the top face) and N (on the bottom face) is called the Hall voltage. This can be measured by use of a voltmeter.

The potential difference measured between the electrodes M and N is given by

$$U_{MN} = U_H + V_{MN}, \tag{1}$$

where U_H is the Hall voltage, V_{MN} is the potential difference in the absence of magnetic field due to some undesired effects (the electrodes M and N being not exactly opposite to each other, etc.).

Normally, the Hall voltage U_H is proportional to $IB \cdot \sin \theta$, and the magnitude of the MRE is proportional to $B^2 \sin^2\theta$, where θ is the angle between vector B and the current direction. But when the sample has a non regular shape, the dependence of U_H and $\dfrac{\Delta R}{R}$ on $B \sin \theta$ may be more complicated.

The Hall effect is used to fabricate a device for measuring the magnetic field. This device is called the Hall sensor. For Hall sensor, the expression of U_H is given by

$$U_H = \alpha \cdot I \cdot B \cdot \sin\theta, \tag{2}$$

where α is, by definition, the sensitivity of the Hall sensor.

II. The measuring sample

The measuring sample in this experiment is a commercial Hall sensor. It consists of a small thin semiconductor plate covered with plastic, with 4 ohmic electrodes, leading to the pins M, N, P, Q (see Fig.5 – 8). It is used in this experiment to study both the MRE and the Hall effect.

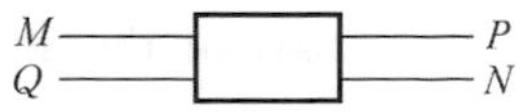

Fig. 5 – 8.

Place the sensor in the magnetic field and use an ohmmeter to mesure the resistance between pins M and N, we can deduce the magnitude of the MRE. Set a current ($I \sim 1$ mA) flowing from P to Q, we can study the Hall effect by measuring the voltage between M and N with a milivoltmeter.

III. Experiment

(1) Determination of the sensitivity α of the Hall sensor.

Set the current through the sensor $I \sim 1$ mA. Keep the distance between the sensor and the centre of the surface of the magnet $y = 2$ cm. Adjust the orientation of magnet for obtaining maximal value of Hall voltage. Measure the Hall voltage with some values of I and determine the sensitivity α of the Hall sensor.

For a magnet having the shape of a disk of radius r, thickness t, the magnetic field at a point situated on its axis at a distance y from the center of the disk surface with $y \gg t$ is given by the expression

$$B(y) = \frac{1}{2} B_0 \left[\frac{y+t}{\sqrt{(y+t)^2 + r^2}} - \frac{y}{\sqrt{y^2 + r^2}} \right], \tag{1}$$

where B_0 is the magnetic induction at the surface of the magnet. The value of B_0 is given on the surface of the magnet.

(2) Study of the dependence of U_H on angle θ between $\boldsymbol{B}$ and the current direction

Set the current through the sensor $I \sim 1$ mA. Keep the distance between the sensor and the centre of the surface of the magnet $y = 2$ cm. Put the magnet on the protractor so that the plane of the magnet is perpendicular to the line connecting the sensor and centre of the magnet.

(a) Draw a sketch of the experimental arrangement.

(b) Tabulate the values of U_H for θ in the range of $-90° \leqslant \theta \leqslant 90°$.

(c) Verify the proportionality between U_H and $\sin \theta$ by using a graph plotted in an appropriate way.

(3) Study of the dependence of $\dfrac{\Delta R}{R}$ on B, for $\boldsymbol{B}$ perpendicular to the sample plane.

The MRE is significant only at sufficiently strong magnetic field. So it is recommended to use a magnetic field as strong as possible.

(a) Draw a sketch of the experimental arrangement and explain the principle of the measurements.

(b) Perform measurements and tabulate the data.

(c) Assume that $\frac{\Delta R}{R} \sim B^k$, determine the value of k by using a graph plotted in an appropriate way. Estimate the maximal deviation of the obtained value of k.

(4) Determination of the relative permeability μ of the ferromagnetic materials of the core of the toroidal coil

Determine the relative permeability μ of the core material at the measured current intensity I by following this guidance step by step:

- Put the Hall sensor into the gap on the core.
- Connect the coil and an ammeter to battery 2. Use only the inputs "COM" and "20 A" of the ammeter in this case.
- Measure the current I in the coil and the magnetic field B in the gap.
- Calculate the value of μ.

You can use the following relation:

$$\frac{B(2\pi\rho - d)}{\mu} + Bd = 4\pi 10^{-7} NI.$$

The average radius of the core $\rho = 25$ mm; $N =$ number of turns; the width of the gap $d = 3$ mm.

Appendix

Instruction for the digital multimeter

(1) Press the Power On/Off button before use.

(2) When the multimeter is used as a milliammeter, the inlets are COM and A. Turn the Function Dial to DCA (2 m to 200 m).

(3) When the multimeter is used as an ammeter, the inlets are COM and 20 A. Turn the Function Dial to DCA (20 m).

(4) When the multimeter is used as a millivoltmeter, the inlets are COM and VΩ. Turn the Function Dial to DCV (200 m to 1000 m).

(5) When the multimeter is used as an ohmmeter, the inlets are COM and VΩ. Turn the Function Dial to Ω (200 m to 200 m).

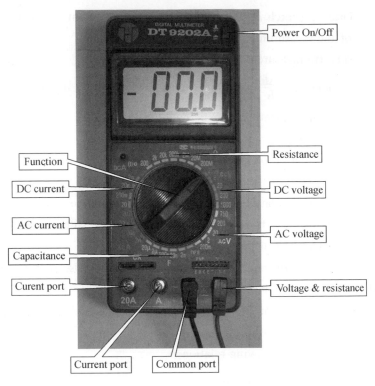

Power On/Off

Function

DC current

AC current

Capacitance

Curent port

Resistance

DC voltage

AC voltage

Voltage & resistance

Current port   Common port

Fig. 5 - 9.

Solution

(1) Determination of the sensitivity α of the Hall sensor

The current I_H supplied by battery 1 and flowing through the miliammeter and the sensor via the pins P and Q is maintained unchanged. Use the variable resistance to maintain $I_H \sim 1$ mA. The Hall voltage U_H is measured across the pins M and N with a milivoltmeter.

In the absence of a magnetic field, we obtain -1.3 mV $< V_{MN} < 1.3$ mV.

Put the magnet on the protractor at a distance $y = 2$ cm $= 2 \times 10^{-2}$ m from the sensor.

Rotate the protractor to obtain the maximum value of U_H. We find the position of the magnet, in which the line joining the magnet and the sensor is perpendicular to the surface of the sensor. In this position of the magnet the axis is perpendicular to the surface of the sensor and B is perpendicular to the current I_H, and $\theta = 90°$.

Measure U_{MN} and calculate the value of U_H: $U_H = U_{MN} - V_{MN}$.

Calculate the value of B by using Eq. (1).

Use the relation $\alpha = \dfrac{U_H}{IB}$ to calculate the value of α.

Repeat the measurement to obtain 4 – 5 different values of α, and calculate the mean value of α and the deviation $\Delta\alpha$.

(2) Study the dependence of U_H on θ.

(a) Keeping y unchanged, $y = 2$ cm. By rotating the protractor, we change the angle θ between B and the current direction.

(b) At each angle θ, measure U_H (always with the same I_H). Vary the angle in the range $-90° \leqslant \theta \leqslant 90°$.

Tabulate U_H versus θ.

For $U_H = \gamma\sin\theta$, then:

$$\ln U_H = \ln\gamma + m \cdot \ln(\sin\theta) \tag{2}$$

Plot $\ln U_H$ versus $\ln(\sin\theta)$, we get m and the error.

We find that the graph of $\ln U_H$ versus $\ln(\sin\theta)$ is a straight line. From the graph we can deduce $m = 1$.

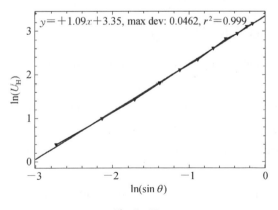

Fig. 5 – 10.

(3) Study the dependence of $\frac{\Delta R}{R}$ on B.

The Hall sensor, with its sensitivity α obtained from part (1), is used to measure the magnetic induction B. It serves to study the magnetoresistivity as well. Because the resistance between N and M does not change during the experiment (and equals about 350 Ω) we can study the dependence ΔR instead of $\frac{\Delta R}{R}$ on B.

(a) Keep the axis of the magnet perpendicular to the surface of the sensor by the procedure described in part (1). Provide a current $I = 1$ mA to the Hall sensor and measure the intensity of the magnetic field (at the sensor).

Turn off the current, and switch the multimeter from the milivoltmeter regime to the ohmmeter regime.

Measure the resistance between N and M and calculate the value of ΔR.

Vary the distance y from the sensor to the magnet from 2 cm to 0.6 cm and repeat the above measurement. Tabulate the values of $\frac{\Delta R}{R}$ or ΔR versus B.

(b) Assuming $\frac{\Delta R}{R} = \beta \cdot B^k$, we have

$$\ln\left(\frac{\Delta R}{R}\right) = \ln \beta + k \ln B. \qquad (1)$$

Plot $\ln\left(\frac{\Delta R}{R}\right)$ versus $\ln B$.

Draw a straight line passing as nearly the experimental points as possible. The slope of this line gives the value of k. One can find the value $k \sim 2$. The deviation of k can be estimated from the graph by using eye-balling method.

The student may use the less square method to determine k and the error.

(4) Determination of the permeability μ of a ferromagnetic core in a toroidal coil.

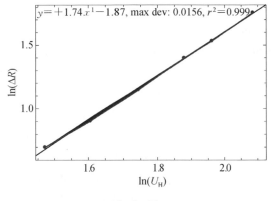

$y = +1.74x^1 - 1.87$, max dev: 0.0156, $r^2 = 0.999$

Fig. 5 – 11.

Put the Hall sensor into the gap. Keep the current through the sensor constant, $I = 1$ mA. The sensor is used to measure B in the gap.

Connect the coil to battery 2 via an ammeter. Note the value I_C of the current flowing in the coil. I_C varies from 3 to 4 amperes. Turn off immediately the current I_C after reading its value, to avoid the variation of the temperature of the sensor.

Count the number N of turns of the coil. From the values of U_H, α, determine B. From the values of B, ρ, d and N, using the relation:

$$\frac{B(\rho - d)}{\mu} + Bd = 4\pi \cdot 10^{-7} \cdot N \cdot I,$$

we obtain the value of μ.

Problem 2
Black Box

Apparatus and materials

(1) A double beam oscilloscope.

(2) A function generator capable to generate sine, triangle and square waves over the 0. 02 Hz to 2 MHz range.

(3) A "Black Box" with two groups of connectors: the $ABCD$ group and $A'B'C'D'$ group. Besides, there are also two connectors for the standard resistor $R_n = 5$ kΩ, which is isolated from the two groups.

(4) Conductors of negligible resistance.

(5) Graph paper.

Warning: You are not allowed to open the black box.

Experiment

In the black box, there are two groups of passive elements (that are elements of the types: resistor R, capacitor C or inductor (induction coil) L). The first group consists of three elements Z_1, Z_2, Z_3 connected in a star circuit as shown in Fig. 5-12. The elements are led out to the connectors A, B, C and D, with A — the common connector of the $ABCD$ group. The second group consists of three elements Z_1', Z_2', Z_3' connected in the same manner to connectors A', B', C' and D', with A' — the common connector of the $A'B'C'D'$ group (see Fig. 5-13).

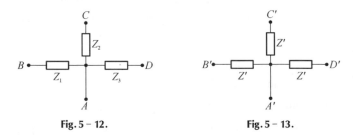

Fig. 5 - 12.   Fig. 5 - 13.

(1) By using the oscilloscope and the function generator, determine the type and the parameter (that is resistance of R, capacity of C, inductivity of L) of each of the elements Z_1, Z_2, Z_3 and Z_1', Z_2', Z_3'.

(2) Connect five points B, C, B', C' and D' together. We obtain a new black box with terminals $DD'A'$ (called $DD'A'$).

(a) Draw the electric circuit of this black box.

(b) Apply a sine wave from the generator to connectors D and A'.

Plot a graph of the ratio of the voltage amplitudes $K = \dfrac{U_{D'A'}}{U_{DA'}}$ and the phase shift φ between these voltages as functions of the frequency f of the signal.

(c) The graphs possess a particular point at a certain frequency f_0. Determine the value of the frequency f_0, the ratio $K = \dfrac{U_{D'A'}}{U_{DA'}}$ and the phase *shift* φ at this frequency.

(d) Derive the relation between f_0 and the parameters of the elements in the black box and calculate the values of f_0.

Appendix

Instruction to instruments

1. Oscilloscope

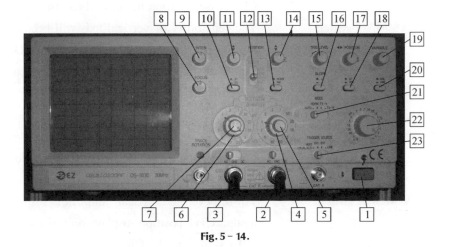

Fig. 5 – 14.

(1) Power ON/OFF Button

(2) CH2 or Y IN connector — For applying an input signal to vertical amplifier CH2, or the Y-axis (Vertical) amplifier during X – Y operation.

(3) CH1 or X IN connector — For applying an input signal to vertical amplifier CH1, or the X-axis (Horizontal) amplifier during X – Y operation.

(4) CH2 VOLTS/DIV switch — To select the calibrated deflection factor of the input signal fed to CH2 vertical amplifier.

(5) (7) VARIABLE controls — Provide continuosly variable adjustement of deflection factor between steps of the VOLTS/DIV switches.

VOLTS/DIV calibration are accurate only when the VARIABLE controls are click-

	stopped in their fully clockwise position.
（6）CH1 VOLTS/DIV switch	To select the calibrated deflection factor of the input signal fed to CH1 vertical amplifier.
（8）FOCUS control	To obtain maximum trace sharpness.
（9）INTEN control	Adjusts the brightness of the CRT display. Clockwise rotation increases brightness.
（10）×5 MAG switch	The sensibility of vertical axis will become 5 times if the switch selected at ×5 MAG. That's to say, the measuring voltage will be 1/5 of indicator value of VOLTS/DIV.
（11）CH1 Position control	For vertically positioning the CH1 trace on the CRT screen, clockwise rotation moves the trace upward, counterclockwise rotation moves the trace down.
（12）V MODE switch	To select the vertical amplifier display mode. CH1 position displays only the CH1 input signal on the CRT screen. CH2 position displays only the CH2 input signal on the CRT screen. DUAL position displays the CH1 and CH2 input signal on the CRT screen.

If trigger Source is selected CH1, CHOP
mode: TIME/DIV 0. 2 s~1 ms
ALT mode: TIME/DIV 0. 5 ms~0. 2 μs
If Trigger Source is selected VERT,
ALT mode: TIME/DIV 0. 2 s~0. 2 μs
ADD position displays the algebraic sum of
CH1 and CH2 signal.

（13）CH2 INV switch	Select switch at INV the signal added to CH2 will be turned over.
（14）CH2 Position control	For vertically positioning the CH2 trace on

the CRT screen, clockwise rotation moves the trace upward, counterclockwise rotation moves the trace down.

(15) Trigger LEVEL control   To select the trigger signal amplitude at which triggering occurs. When rotated clockwise, the trigger point moves toward the positive peak of the trigger signal. When this control is rotated counterclockwise, the trigger point moves toward the negative peak of the trigger signal.

(16) Trigger Slope switch   To select the positive or negative slope of the trigger signal (on LEVEL control) for initiating sweep. Pulled in, the switch selects the positive (+) slope. When pushed, this switch selects the negative (−) slope.

(17) Horizontal Position control   To adjust the horizontal position of the traces displayed on control the CRT. Clockwise rotation moves the traces to the right, counterclockwise rotation moves the traces to the left.

(18) ×10 MAG switch   Placing the switch on × 10 MAG sweep time will be expanded to 10 times and in this instance sweep time becomes 1/10 of TIME/DIV indicator value.

(19) VARIABLE control   Provides continuously variable adjustment of sweep rate between steps of the TIME/DIV switch. TIME/DIV calibrations are accurate only when the VARIABLE control is click-stopped fully clockwise.

(20) VARIABLE control

(21) Trigger MODE switch   To select the sweep triggering mode.

AUTO position selects free-running sweep where a baseline is displayed in the absence of a signal.

This condition automatically reverts to triggered sweep when a trigger signal of 25 Hz or higher is received and other trigger controls are properly set.

NORM position produces sweep only when a trigger signal is received and other controls are properly set. No trace is visible when the signal frequency is 25 Hz or lower.

TV – V and TV – H positions are used for observing composite video signals. They are not used in our case.

(22) TIME/DIV switch   To select either the calibrated sweep rate of the main timebase, the delay time range for delayed sweep operation or X – Y operation.

(23) Trigger Source switch   To conveniently select the trigger source. In our case, set it to CH1.

 2. Function generator

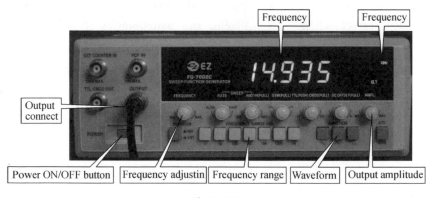

Fig. 5 – 15.

• The Power button may be pressed for "ON" and pressed again for

"OFF".

  • Select the Frequency Range and press the proper button.

  • The frequency is shown on the digital display.

  • Use the Frequency adjusting knob to tune to the proper frequency.

  • Select the Waveform (sine wave, triangle wave or square wave) by pressing the appropriate button.

Solution

(1) The type of the elements Z.

Adjust the oscilloscope to obtain the same gain for the two channels. Use the sine wave from the generator.

Use the circuit in Fig. 5 – 16 to determine the type of the elements.

We find finally:

Z_3, Z'_1 and Z'_2 are resistors and $Z'_1 = Z'_2 = 2Z_3 = (10.0 \pm 0.5)$ kΩ,

Z_1, Z_2 and Z'_3 are capacitors, and $C_1 = C_2 = \dfrac{1}{2}C'_3 = (47 \pm 2)$ nF.

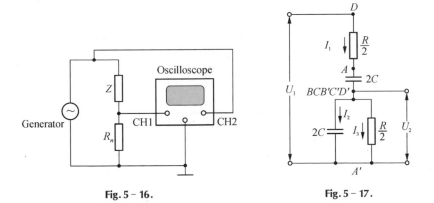

Fig. 5 – 16. **Fig. 5 – 17.**

(2) (a) The electric circuit of the black box $DD'A'$ is shown in Fig. 5 – 17.

(b) Apply a sine wave signal to connectors D and A'. Connect D and A' to channel 1 (CH1) and D' and A' to channel 2 (CH2).

The ratio $K = \dfrac{U_{D'A'}}{U_{DA'}}$ is determined from the amplitudes of the signals

$U_{D'A'}$ và $U_{DA'}$. The phase shift φ can be determined directly from the traces of the signals (or from the Lissajous patterns).

Tabulating K and φ versus f, we get for example:

f	100	150	200	300	330	400	600	800
K		0.29	0.30	0.32	0.33	0.32	0.3	0.28
φ	44	28	16	2	0	-8	-24	-38

The experimental error of f is ± 1 Hz, of φ is $\pm 5°$ and of K is ± 0.02. Here are the plots of K and φ as functions of f.

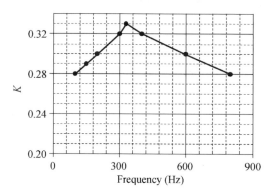

Fig. 5 - 18.

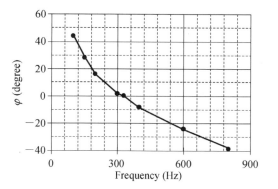

Fig. 5 - 19.

(c) The graphs possess a particular point at $f_0 = 330 \pm 1$ Hz, at which φ equals zero and K has a maximum of $K = 0.33 \pm 0.01$. φ changes its sign from positive to negative with increasing of the frequency f across f_0.

The value of f_0 may vary from 325 Hz to 335 Hz depending on the set of experiment, due to the deviation in the value of the resistance and capacitance in the set.

(d) The phasor diagram for the circuit is shown in Fig. 5 – 20, where u_1 is the instantaneous voltage between D and D', and U_1— its amplitude.

We have $\tan \alpha_1 = \dfrac{1}{\omega CR}$ and $U_1 = I_1 \sqrt{\left(\dfrac{R}{2}\right)^2 + \dfrac{1}{4C^2 \omega^2}}$.

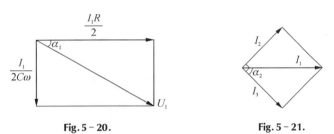

Fig. 5 – 20.   **Fig. 5 – 21.**

For the $D'A'$ parallel cicuit, $I_1 = I_2 + I_3$, and the phasor diagram is shown in Fig. 5 – 21. U_2 is in phase with I_3.

$\tan \alpha_2 = \dfrac{I_2}{I_3} = \omega \cdot 2C \cdot \dfrac{R}{2} = \omega CR$. Let U_2 the voltage between D' and A',

we have $I_3 = \dfrac{U_2}{\dfrac{R}{2}}$; $I_2 = U_2 \cdot 2C \cdot \omega$. Hence

$$U_2 = I_1 \cdot \frac{1}{\sqrt{\dfrac{1}{\left(\dfrac{R}{2}\right)^2} + 4C^2 \omega^2}}.$$

By combining Fig. 5 – 20 and Fig. 5 – 21, we obtain Fig. 5 – 22, with $U = U_1 + U_2$ being the instantaneous voltage between D and A'.

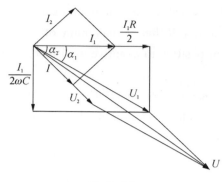

Fig. 5 – 22.

For $\omega = 2\pi f = \dfrac{1}{CR}$, $\tan\alpha_1 = \dfrac{1}{\omega CR} = \tan\alpha_2 = \omega CR$. In this condition,

U_1, U_2 and U are in phase, so that $\varphi = \alpha_2 - \alpha_1 = 0$. Hence $K = \dfrac{U_{D'A'}}{U_{DA'}} =$

$\dfrac{U_2}{U_1 + U_2}$. Substituing $\omega = \dfrac{1}{CR}$ into U_1 and U_2, we obtain $K = \dfrac{1}{3}$.

That is, for $f_0 = \dfrac{\omega}{2\pi} = \dfrac{1}{2\pi RC} = \dfrac{1}{2\pi \cdot 10^4 \cdot 48 \cdot 10^{-9}} = 331$ Hz, $K =$

$\dfrac{1}{3} = 0.33$, $\varphi = 0$, which is observed in the experiment.

For $\omega \neq \dfrac{1}{CR}$, U_1 and U_2 are out of phase, and K has values smaller than

$\dfrac{1}{3}$.

Minutes of the Sixth Asian Physics Olympiad

Pekanbaru (Riau, Indonesia), April 24 – May 2, 2005

1. The following 16 countries were present at the 6th Asian Physics Olympiad: Azerbaijan (6 students + 2 leaders), Cambodia (8 students + 2 leaders + 3 observers), China (8 students + 2 leaders + 1 observer), Indonesia (8 students + 2 leaders), Jordan (7 students + 2 leaders), Kazakhstan (8 students + 2 leaders + 2 observers), Kyrgyzstan (2 students + 1 leader), Laos PDR (2 students + 1 leader), Malaysia (8 students + 2 leaders + 1 observer), the Philippines (2 students + 1 leader), Qatar (6 students + 2 leaders), Singapore (8 students + 2 leaders), Chinese Taipei (8 students + 2 leaders + 1 observer), Tajikistan (2 students + 1 leader), Thailand (8 students + 2 leaders + 1 observer) and Vietnam (8 students + 2 leaders + 3 observers).

Two countries were present as observers, i.e. the United Arab Emirates (2 persons) and Turkmenistan (1 person).

Indonesia was represented by the regular team and by two guest teams: Indonesia B (4 students + 1 leader) and Indonesia C (8 students from Riau + 2 leaders).

In addition to the above 16 countries a team from Russia (7 students and two leaders) participated as a guest team.

This year Azerbaijan, Qatar and Tajikistan participated in the APhO for the first time.

2. The Opening Ceremony was honoured by the presence of His Excellency Vice-President of the Republic of Indonesia Mr. Jusuf Kalla. At both Opening and Closing Ceremonies His Excellency the Minister of National Education of the Republic of Indonesia, Prof. Dr. Bambang Sudibyo and His Excellency Mr. Rusli Zainal, the Governor of Riau were present.

3. Results of marking the papers by the organisers were presented as the following:

The best score (45.60 points) was achieved by Li Fang from China (the Absolute Winner of the 6th APhO). The second score (45.15 points) by Han-Hsuan Lin from Chinese Taipei. The third score Ming-Jie Dai (44.60 points) from China. The following limits for awarding the medals and the honourable mention were established according to the Statutes.

Gold Medal.	40 points,
Silver Medal.	35 points,
Bronze Medal.	29 points,
Honourable Mention.	22 points.

According to the above limits 14 Gold Medals, 6 Silver Medals, 13 Bronze Medals and 19 Honourable Mentions were awarded. The list of the scores of the winners and the students awarded with honourable mentions were distributed to all the delegations.

4. In addition to the regular prizes the following special prizes were awarded.

● for the Absolute Winner with 45. 60 points. Li Fang (China)

● for the best result in the theoretical part of the competition with 29. 00 points. Yin-Hsuan Lin (Chinese Taipei)

● for the best result in the experimental part of the competition with 19. 15 points. Han-Hsuan Lin (Chinese Taipei)

● for the best female participant. Charmaine Sia Jia Min (Singapore)

● for the best participant in the foreign guest team. Ivan Kiselev (Russia)

● for the best participants from each new country (Prize created by the President of APhOs)[i].

Azer Eyvazov	(Azerbaijan)
Mukhtordzhon Soliev	(Tajikistan)
Turki Obaid N. M. Jamaan	(Qatar)

● for the best participant from the host Province (Riau). Bambang Dwiputra (Indonesia C)

● for the youngest participant. Alexander Afanas'ev (Russia)

5. Prof. Yohanes Surya presented the last version of the list of the organizers of the next APhOs.

7th	2006	Almaty	Kazakhstan	invitation done
8th	2007	not decided yet	China	not reconfirmed yet
9th	2008	not decided yet	Australia	preliminary contacts
10th	2009	not decided yet	Uzbekistan	
11th	2010	not decided yet	Malaysia	preliminary contacts
12th	2011	not decided yet	Israel	confirmed
13th	2012	not decided yet	Chinese Taipei	preliminary contacts

14th   2013   not decided yet   Cambodia   preliminary contacts

As the Secretariat of the APhOs has lost contacts with Uzbekistan, certain changes in the above table are possible.

6. The International Board has unanimously re-elected Prof. Yohanes Surya to the post of President of the Asian Physics Olympiads for the next five years, i.e. to May 2010. The re-election was welcomed by applause.

7. Acting on behalf of the organisers of the next Asian Physics Olympiad in Kazakhstan, the Kazakhi delegation announced that the 7th Asian Physics Olympiad would be organised in Almaty (Kazakhstan) from April 23rd to May 1st, 2006 and cordially invited all the participating countries to attend the competition. Appropriate brochure was disseminated to all the delegation leaders and observers.

8. Dr. Waldemar Gorzkowski (Honorary President of the APhOs), acting on behalf of all the Members of the International Board, expressed deep thanks to the Organizers and all other people involved in the works at the Olympiad in Pekanbaru for excellent organisation and execution of the 6th APhO.

Pekanbaru, January 5, 2005

Dr. Rachmat Widodo Adi	**Prof. Ming-Juey Lin**	**Prof. Yohanes Surya**
Chairman of the 6th APhO's Organizing Committee	Secretary of the APhOs	President of the APhOs

---

[i] For a mistake not all of these Prizes were mentioned at the Closing Ceremony. The missing diplomas were sent to the students by regular post.

Theoretical Competition

April 27,2005   Time available: 5 hours

Problem 1A
Spring Cylinder With Massive Piston

Consider $n = 2$ moles of ideal Helium gas at a pressure P_0, volume V_0 and temperature $T_0 =$ 300 K placed in a vertical cylindrical container (see Fig. 6 – 1). A moveable frictionless horizontal piston of mass $m = 10$ kg (assume $g = 9.8$ m/s^2) and cross section $A = 500$ cm^2 compresses the gas leaving the upper section of the container void. There is a vertical spring attached to the piston and the upper wall of the container. Disregard any gas

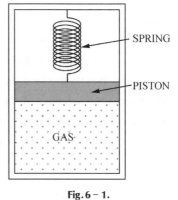

Fig. 6 – 1.

leakage through their surface contact, and neglect the specific thermal capacities of the container, piston and spring. Initially the system is in equilibrium and the spring is unstretched. Neglect the spring's mass.

(a) Calculate the frequency f of small oscillation of the piston, when it is slightly displaced from equilibrium position.

(b) Then the piston is pushed down until the gas volume is halved, and released with zero velocity. Calculate the value(s) of the gas volume when the piston speed is $\sqrt{\dfrac{4gV_0}{5A}}$.

Let the spring constant $k = \dfrac{mgA}{V_0}$. All the processes in gas are adiabatic. Gas constant $R = 8.314$ JK^{-1}mol^{-1}. For mono-atomic gas (Helium) use Laplace constant $\gamma = \dfrac{5}{3}$.

Solution

(a) Gas Volume

At the initial condition, the system is in equilibrium and the spring is unstreched; therefore

$$P_0 A = mg \quad \text{or} \quad P_0 = \frac{mg}{A}. \tag{1}$$

The initial volume of gas

$$V_0 = \frac{nRT_0}{P_0} = \frac{nRT_0 A}{mg}. \tag{2}$$

The work done by the gas from $\frac{1}{2} V_0$ to V

$$W_{\text{gas}} = \int_{\frac{V_0}{2}}^{V} P dV = \int_{\frac{V_0}{2}}^{V} \frac{P_0 V_0^{\gamma}}{V^{\gamma}} dV = \frac{P_0 V_0^{\gamma}}{1 - \gamma} \left[V^{1-\gamma} - \left(\frac{V_0}{2} \right)^{1-\gamma} \right]. \tag{3}$$

Eq. (3) can also be obtained by calculating the internal energy change (without integration)

$$W_{\text{gas}} = -\Delta E = -nC_V (T - T'_0), \tag{4}$$

where T'_0 is the temperature when the gas volume is $\frac{V_0}{2}$.

The change of the gravitational potential energy

$$\Delta E_p = mg \Delta h = mg \frac{V - \frac{1}{2} V_0}{A}. \tag{5}$$

The change of the potential energy of the spring

$$
\begin{aligned}
\Delta_{\text{spring}} &= \frac{1}{2} kx^2 - \frac{1}{2} kx_0^2 \\
&= \frac{1}{2} \left(\frac{mgA}{V_0} \right) \left(\frac{V_0 - V}{A} \right)^2 - \frac{1}{2} \left(\frac{mgA}{V_0} \right) \left[\frac{V_0 - \frac{V_0}{2}}{A} \right]^2 \\
&= \frac{1}{2} \frac{mgV_0}{A} \left(1 - \frac{V}{V_0} \right)^2 - \frac{1}{8} \left(\frac{mgV_0}{A} \right).
\end{aligned} \tag{6}
$$

The kinetic energy

$$E_k = \frac{1}{2}mv^2 = \frac{1}{2}m\frac{4gV_0}{5A} = \frac{2mgV_0}{5A}. \tag{7}$$

By conservation of energy, we have

$$W_{gas} = \Delta E_p + \Delta_{spring} + E_k, \tag{8}$$

$$\frac{P_0 V_0^\gamma}{1-\gamma}\left[V^{1-\gamma} - \left(\frac{V_0}{2}\right)^{1-\gamma}\right] = mg\,\frac{V - \dfrac{V_0}{2}}{A} + \frac{1}{2}\frac{mgV_0}{A}\left(1 - \frac{V}{V_0}\right)^2$$

$$-\frac{1}{8}\frac{mgV_0}{A} + \frac{2}{5}\frac{mgV_0}{A}, \tag{9}$$

$$\frac{mgV_0}{A(1-\gamma)}\left[\frac{V^{1-\gamma}}{V_0^{1-\gamma}} - \left(\frac{1}{2}\right)^{1-\gamma}\right] = mg\,\frac{V - \dfrac{V_0}{2}}{A} + \frac{mgV_0}{2A}\left(1 - \frac{V}{V_0}\right)^2$$

$$+\frac{11}{40}\frac{mgV_0}{A}. \tag{10}$$

Let $s = \dfrac{V}{V_0}$ so the above equation becomes

$$\frac{1}{1-\gamma}\left[s^{1-\gamma} - \left(\frac{1}{2}\right)^{1-\gamma}\right] = \left(s - \frac{1}{2}\right) + \frac{1}{2}(1-s)^2 + \frac{11}{40}. \tag{11}$$

With $\gamma = \dfrac{5}{3}$ we get

$$0 = \frac{1}{2}s^2 + \frac{11}{40} + \frac{3}{2}\left[s^{-\frac{2}{3}} - \left(\frac{1}{2}\right)^{-\frac{2}{3}}\right]. \tag{12}$$

Solving Eq. (12) numerically, we get

$$s_1 = 0.74 \text{ and } s_2 = 1.30.$$

Therefore $V_1 = 0.74V_0 = 0.74\,\dfrac{nRT_0A}{mg} = 1.88 \text{ m}^3$ *or* $V_2 = 1.30V_0 =$

3.31 m^3.

   (b) Small Oscillation

   The equation of motion when the piston is displaced by x from the equilibrium position is

$$m\ddot{x} = -kx - PA + mg,\qquad(13)$$

P is the gas pressure

$$P = \frac{P_0 V_0^\gamma}{V^\gamma} = \frac{P_0 V_0^\gamma}{(V_0 - Ax)^\gamma}$$

$$= \frac{P_0}{\left(1 - \dfrac{Ax}{V_0}\right)^\gamma}.\qquad(14)$$

Since $Ax \ll V_0$ then we have $P \approx P_0\left(1 + \gamma\dfrac{Ax}{V_0}\right)$, therefore

$$m\ddot{x} \approx -kx - P_0 A\left(1 + \gamma\frac{Ax}{V_0}\right) + mg,$$

$$m\ddot{x} \approx -\left[k + P_0 A\left(\gamma\frac{A}{V_0}\right)\right]x,\qquad(15)$$

$$m\ddot{x} \approx -\left[\frac{mgA}{V_0} + \frac{mg}{A}A\left(\gamma\frac{A}{V_0}\right)\right]x,$$

$$m\ddot{x} + (1 + \gamma)\frac{mgA}{V_0}x \approx 0.$$

The frequency of the small oscillation is

$$f = \frac{1}{2\pi}\sqrt{(1 + \gamma)\frac{gA}{V_0}}$$

$$= \frac{1}{2\pi}\sqrt{(1 + \gamma)\frac{mg^2}{nRT_0}}.\qquad(16)$$

Numerically $f = 0.114$ Hz.

Problem 1B
The Parametric Swing

A child builds up the motion of a swing by standing and squatting. The trajectory followed by the center of mass of the child is illustrated in Fig. 6 – 2. Let r_u be the radial distance from the swing pivot to the child's center of mass when the child is standing, while r_d is the radial distance from the swing pivot to the child's center of mass when the child is squatting. Let the ratio of r_d to r_u be $2^{\frac{1}{10}} = 1.072$, that is the child moves its center of mass by

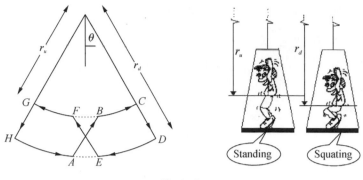

Fig. 6 – 2.

roughly 7% compared to its average radial distance from the swing pivot.

To keep the analysis simple it is assumed that the swing be mass-less, the swing amplitude is sufficiently small and that the mass of the child resides at its center of mass. It is also assumed that the transitions from squatting to standing (the A to B and the E to F transitions) are fast compared to the swing cycle and can be taken to be instantaneous. It is similarly assumed that the squatting transitions (the C to D and the G to H transitions) can also be regarded as occurring instantaneously.

How many cycles of this maneuver does it take for the child to build up the amplitude (or the maximum angular velocity) of the swing by a factor of two?

Solution

(1) The conservation of angular momentum (CAM) from A to B, C to D, E to F and G to H.

$$L = I\boldsymbol{\theta} = mr^2\,\boldsymbol{\theta}, \tag{1}$$

m = mass of the child,

r = distance of the child's center of mass to the swing's pivot P,

$\boldsymbol{\theta}$ = the swing's angular velocity with respect to P.

A to B:

Let $\boldsymbol{\theta}_d$ and $\boldsymbol{\theta}_u$ are the angular velocity at point A and B respectively, then according to CAM,

$$L_A = mr_d^2\, \boldsymbol{\theta}_d = L_B = mr_u^2\, \boldsymbol{\theta}_u,$$ (2)

so that

$$\boldsymbol{\theta}_d = \frac{r_u^2}{r_d^2}\, \boldsymbol{\theta}_u,$$ (3)

hence each time the swing repeat moving upward (A to B or E to F) its angular speed increases by factor of $\left(\frac{r_d}{r_u}\right)^2$.

(2) The Conservation of Mechanical Energy (from B to C)

$$E_B = E_C = K + V = \frac{1}{2}mr_u^2\, \boldsymbol{\theta}_B^2 - mgr_u(1 - \cos\theta).$$ (4)

The change of the potential energy (from B to C) is the same as the rotation energy at point B,

$$mgr_u(1 - \cos\theta) = \frac{1}{2}mr_u^2\, \boldsymbol{\theta}_u^2.$$ (5)

Using the similar method, we could get the following equation for the transition from D to E,

$$mgr_d(1 - \cos\theta) = \frac{1}{2}mr_d^2\, \boldsymbol{\theta}_d^2.$$

From equations (3), (5) and (6) we have

$$\frac{r_u}{r_d} = \left(\frac{r_u}{r_d}\right)^2\left(\frac{\boldsymbol{\theta}_u}{\boldsymbol{\theta}_d}\right)^2 \rightarrow \frac{\boldsymbol{\theta}_{d'}}{\boldsymbol{\theta}_u} = \sqrt{\frac{r_u}{r_d}}.$$ (7)

For half a cycle we have $\boldsymbol{\theta}_{u'} = \left(\frac{r_d}{r_u}\right)^2\boldsymbol{\theta}_{d'} = \left(\frac{r_d}{r_u}\right)^{\frac{3}{2}}\boldsymbol{\theta}_u.$ (8)

For n complete cycles, the growth of angular velocity amplitude as well as the angular amplitude θ_A increases by a factor of $\rho_{A,\,n} = \left(\frac{r_d}{r_u}\right)^{3n}.$

For $\rho_{A,\,n} = 2$ then with $\frac{r_d}{r_u} = 2^{\frac{1}{10}}$ one gets $(2^{\frac{1}{10}})^{3n} = 2$, so $n = \frac{10}{3}.$

Alternate Solution

The moment of inertia with respect to the swing pivot

$$I = Mr^2. \tag{1}$$

Since the A to B transition is fast one has by conservation of angular momentum,

$$I_A \omega_A = I_B \omega_B. \tag{2}$$

The energy at point A is

$$E_A = \frac{1}{2} I_A \omega_A^2. \tag{3}$$

The energy at point B is

$$E_B = \frac{1}{2} I_B \omega_B^2 + Mgh, \tag{4}$$

where $h = r_d - r_u$ is the vertical distance the child's center of mass moves. The energy at point C (conservation of energy)

$$E_C = E_B = \frac{1}{2} I_B \omega_B^2 + Mgh. \tag{5}$$

As the child squats at the C to D transition, the swing losses energy of the amount Mgh, so

$$E_D = \frac{1}{2} I_B \omega_B^2. \tag{6}$$

Energy at point E is equal to energy at point D (conservation energy)

$$E_E = E_D = \frac{1}{2} I_B \omega_B^2. \tag{7}$$

But we have also

$$E_E = \frac{1}{2} I_E \omega_E^2. \tag{8}$$

From equation (7) and (8) we have

$$\omega_E^2 = \frac{I_B}{I_E} \omega_B^2. \tag{9}$$

Using equation (2) this equation yields

$$\omega_E^2 = \frac{I_A}{I_B}\omega_A^2,\tag{10}$$

where we have used $I_E = I_A$.

Using equation (1) one obtains from equation (10)

$$\omega_E^2 = \frac{r_d^2}{r_u^2}\omega_A^2.\tag{11}$$

From this one obtains

$$\frac{|\omega_E|}{|\omega_A|} = \frac{r_d}{r_u}.\tag{12}$$

This ratio gives the fractional increase in the amplitude for one half cycle of the swing motion. The fractional increase in the amplitude after n cycles is thus,

$$\frac{|\omega_E|_n}{|\omega_A|_0} = \left(\frac{r_d}{r_u}\right)^{2n},\tag{13}$$

where $|\omega_A|_0$ is the initial amplitude and $|\omega_E|_n$ is the amplitude after n cycles. Substitute the values,

$$2 = 2^{\frac{2n}{10}}\tag{14}$$

or,

$$n = 5.\tag{15}$$

Thus it takes only 5 swing cycles for the amplitude to build up by a factor of two.

Problem 2
Magnetic Focusing

There exist many devices that utilize fine beams of charged particles. The cathode ray tube used in oscilloscopes, in television receivers or in electron microscopes. In these devices the particle beam is focused and deflected in much the same manner as a light beam is in an optical instrument.

Beams of particles can be focused by electric fields or by magnetic fields. In problem 2A and 2B we are going to see how the beam can be focused by a magnetic field.

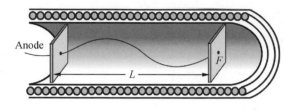

Fig. 6 – 3.

2A. Magnetic Focusing Solenoid

Fig. 6 – 3 shows an electron gun situated inside (near the middle) a long solenoid. The electrons emerging from the hole on the anode have a small transverse velocity component. The electron will follow a helical path. After one complete turn, the electron will return to the axis. By adjusting the magnetic field B inside the solenoid correctly, all the electrons will converge at the same point F after one complete turn. Use the following data:

- The voltage difference that accelerates the electrons $V = 10$ kV.
- The distance between the anode and the focus point F, $L = 0.5$ m.
- The mass of an electron $m = 9.11 \times 10^{-31}$ kg.
- The charge of an electron $e = 1.6 \times 10^{-19}$ C.
- $\mu_0 = 4\pi \times 10^{-7}$ H/m.
- Treat the problem non-relativistically.

(a) Calculate B so that the electron returns to the axis at point F after one complete turn.

(b) Find the current in the solenoid if the latter has 500 turns per meter.

Solution

(a) In magnetic field, the particle will be deflected and follow a helical path.

Lorentz Force in a magnetic field B,

$$\frac{mv_\perp^2}{R} = ev_\perp B, \tag{1}$$

where $v_\perp$ is the transverse velocity of the electron, R is the radius of the path.

Since $v_\perp = \omega R$ $\left(\omega = \dfrac{2\pi}{T}\right.$ is the particle angular velocity and T is the period$\Big)$, then,

$$m\frac{2\pi}{T} = eB. \tag{2}$$

To be focused, the period of electron T must be equal to $\dfrac{L}{v_/}$, where $v_/$ is the parallel component of the velocity.

We also know,

$$eV = \frac{1}{2}m(v_\perp^2 + v_/^2) \approx \frac{1}{2}mv_/^2. \tag{3}$$

All the information above leads to

$$B = 2^{\frac{3}{2}}\pi\frac{\left(\dfrac{mV}{e}\right)^{\frac{1}{2}}}{L}. \tag{4}$$

Numerically

$$B = 4.24 \text{ mT}.$$

(b) The magnetic field of the Solenoid:

$$B = \mu_0 In, \tag{5}$$

$$I = \frac{B}{n\mu_0}. \tag{6}$$

Numerically

$$I = 6.75 \text{ A}.$$

2B. Magnetic Focusing (Fringing Field)

Two pole magnets positioned on horizontal planes are separated by a certain distance such that the magnetic field between them be B in vertical direction (see Fig. 6 − 4). The poles faces are rectangular with length l and width w. Consider the fringe field near the edges of the poles (fringe field is field particularly associated to the edge effects). Suppose the extent of the fringe field is b (see Fig. 6 − 5). The fringe field has two components $B_x \boldsymbol{i}$ and $B_z \boldsymbol{k}$. For simplicity assume that $|B_x| = \dfrac{B|z|}{b}$ where $z = 0$ is the mid plane of the gap, explicitly:

when the particle enters the fringe field $B_x = + B\dfrac{z}{b}$,

when the particle enters the fringe field after traveling through the magnet, $B_x = - B\dfrac{z}{b}$.

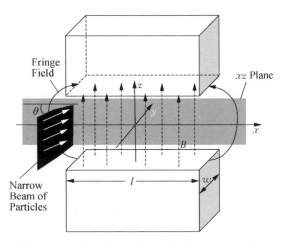

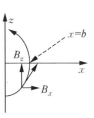

Fig. 6 − 4.   Overall view (note that θ is very samll).   **Fig. 6 − 5.**   Fringe field.

A parallel narrow beam of particles, each of mass m and positive charge q enters the magnet (near the center) with a high velocity v parallel to the horizontal plane. The vertical size of the beam is comparable to the distance between the magnet poles. A certain beam enters the magnet at an angle θ from the center line of the magnet and leaves the magnet at an angle $- \theta$ (see Fig. 6 − 6. Assume θ is very small). Assume that the angle θ with which the particle enters the fringe field is the same as the angle θ when it enters the uniform field.

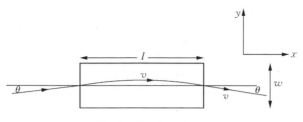

Fig. 6 – 6. Top view.

The beam will be focused due to the fringe field. Calculate the approximate focal length if we define the focal length as illustrated in Fig. 6 – 7 (assume $b \ll l$ and assume that the z-component of the deflection in the uniform magnetic field B is very small).

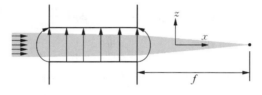

Fig. 6 – 7. Side view.

Solution

The magnetic force due to the fringe field on charge q with velocity v is

$$\mathbf{F} = q\mathbf{v} \times \mathbf{B}. \tag{1}$$

The z-component of the force obtained from the cross product is

$$F_z = q(v_x B_y - v_y B_x) = -qv_y B_x = -\frac{qvB_z \sin\theta}{b}. \tag{2}$$

The vertical momentum gained by the particle after entering the fringe field

$$\Delta P_z = \int F_z dt = -\frac{qvB_z \sin\theta}{b}\Delta t$$

$$= -\frac{qvB_z \sin\theta}{b}\frac{b}{v\cos\theta} = -qB_z \tan\theta. \tag{3}$$

The particle undergoes a circular motion in the constant magnetic field

B region

$$m \frac{v^2}{R} = qvB , \tag{4}$$

$$v = \frac{qBR}{m} = \frac{qBl}{2m\sin \theta}. \tag{5}$$

Therefore,

$$\sin \theta = \frac{qBl}{2mv}. \tag{6}$$

After the particle exits the fringe field at the other end, it will gain the same momentum.

The total vertical momentum gained by the particle is

$$(\Delta P_z)_{\text{total}} = 2\Delta P_z = -2qB_z \tan \theta \approx -2qB_z \frac{qBl}{2mv} = -\frac{q^2 B_z^2 l}{mv}. \tag{7}$$

Note that for small θ, we can approximate $\tan \theta \approx \sin \theta$.

Meanwhile, the momentum along the horizontal plane (xy-plane) is

$$p = mv. \tag{8}$$

From the geometry in Fig. 6 – 6, we can get the focal length by the following relation,

$$\frac{|\Delta P_z|}{p} = \frac{|Z|}{f}, \tag{9}$$

$$f = \frac{m^2 v^2}{q^2 B^2 l}.$$

Problem 3
Light Deflection by a Moving Mirror

Reflection of light by a relativistically moving mirror is not theoretically new. Einstein discussed the possibility or worked out the process using the Lorentz transformation to get the reflection formula due to a mirror moving with a velocity v. This formula, however, could also be derived by using a relatively simpler method. Consider the reflection process as shown in Fig. 6 – 8, where a plane mirror M moves with a velocity $\mathbf{v} = v\mathbf{e}_x$ (where $\mathbf{e}_x$

is a unit vector in the x-direction) observed from the lab frame F. The mirror forms an angle ϕ with respect to the velocity (note that $\phi \leqslant 90°$, see Fig. 6 – 8). The plane of the mirror has n as its normal. The light beam has an incident angle α and reflection angle β which are the angles between n and the incident beam 1 and reflection beam $1'$, respectively in the laboratory frame F. It can be shown that

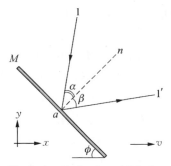

Fig. 6 – 8. Reflection of light by a relativistically moving mirror.

$$\sin \alpha - \sin \beta = \frac{v}{c} \sin \phi \sin(\alpha + \beta). \tag{1}$$

3A. Einstein's Mirror

About a century ago Einstein derived the law of reflection of an electromagnetic wave by a mirror moving with a constant velocity $v = -v\boldsymbol{e}_x$ (see Fig. 6 – 9). By applying the Lorentz transformation to the result obtained in the rest frame of the mirror, Einstein found that:

$$\cos \beta = \frac{\left[1 + \left(\frac{v}{c}\right)^2\right]\cos \alpha - 2\frac{v}{c}}{1 - 2\frac{v}{c}\cos \alpha + \left(\frac{v}{c}\right)^2}. \tag{2}$$

Derive this formula using Eq. (1) without Lorentz transformation!

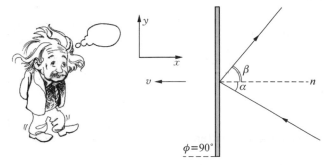

Fig. 6 – 9. Einstein mirror moving to the left with a velocity v.

3B. Frequency Shift

In the same situation as in 3A, if the incident light is a monochromatic beam hitting M with a frequency f, find the new frequency f' after it is reflected from the surface of the moving mirror. If $\alpha = 30°$ and $v = 0.6\,c$ in Fig. 6 – 9, find frequency shift Δf in percentage of f.

3C. Moving Mirror Equation

Fig. 6 – 10 shows the positions of the mirror at time t_0 and t. Since the observer is moving to the left, the mirror moves relatively to the right. Light beam 1 falls on point a at t_0 and is reflected as beam $1'$. Light beam 2 falls on point d at t and is reflected as beam $2'$. Therefore, $\overline{ab}$ is the wave front of the incoming light at time t_0. The atoms at point are disturbed by the incident wave front $\overline{ab}$ and begin to radiate a wavelet. The disturbance due

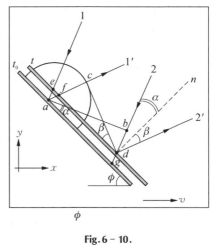

Fig. 6 – 10.

to the wave front $\overline{ab}$ stops at time t when the wavefront strikes point d.

By referring to Fig. 6 – 10 for light wave propagation or using other methods, derive Eq. (1).

🔑 Solution

3A. Einstein's Mirror

By taking $\phi = \dfrac{\pi}{2}$ and replacing v with $-v$ in Eq. (1), we obtain

$$\sin\alpha - \sin\beta = -\frac{v}{c}\sin(\alpha+\beta). \tag{3}$$

This equation can also be written in the form of

$$\left(1 + \frac{v}{c}\cos\beta\right)\sin\alpha = \left(1 - \frac{v}{c}\cos\alpha\right)\sin\beta. \tag{4}$$

The square of this equation can be written in terms of a squared equation of

$\cos \beta$, as follows,

$$\left(1-2\frac{v}{c}\cos\alpha+\frac{v^2}{c^2}\right)\cos^2\beta+2\frac{v}{c}(1-\cos^2\alpha)\cos\beta$$

$$+2\frac{v}{c}\cos\alpha-\left(1+\frac{v^2}{c^2}\right)\cos^2\alpha=0, \tag{5}$$

which has two solutions,

$$(\cos\beta)_1=\frac{2\dfrac{v}{c}\cos^2\alpha-\left(1+\dfrac{v^2}{c^2}\right)\cos\alpha}{1-2\dfrac{v}{c}\cos\alpha+\dfrac{v^2}{c^2}} \tag{6}$$

and

$$(\cos\beta)_2=\frac{-2\dfrac{v}{c}+\left(1+\dfrac{v^2}{c^2}\right)\cos\alpha}{1-2\dfrac{v}{c}\cos\alpha+\dfrac{v^2}{c^2}}. \tag{7}$$

However, if the mirror is at rest ($v=0$) then $\cos\alpha=\cos\beta$; therefore the proper solution is

$$\cos\beta_2=\frac{-2\dfrac{v}{c}+\left(1+\dfrac{v^2}{c^2}\right)\cos\alpha}{1-2\dfrac{v}{c}\cos\alpha+\dfrac{v^2}{c^2}}. \tag{8}$$

3B. Frequency Shift

The reflection phenomenon can be considered as a collision of the mirror with a beam of photons each carrying an incident and reflected momentum of magnitude

$$p_f=\frac{hf}{c} \text{ and } p_f{}'=\frac{hf'}{c}. \tag{9}$$

The conservation of linear momentum during its reflection from the mirror for the component parallel to the mirror appears as

$$p_f\sin\alpha=p_f{}'\sin\beta \text{ or } f'\sin\beta$$

$$=f'\frac{\left(1-\dfrac{v^2}{c^2}\right)\sin\alpha}{\left(1+\dfrac{v^2}{c^2}\right)-2\dfrac{v}{c}\cos\alpha}$$

$$= f\sin \alpha. \tag{10}$$

Thus

$$f' = \frac{\left(1+\frac{v^2}{c^2}\right)-2\frac{v}{c}\cos \alpha}{\left(1-\frac{v^2}{c^2}\right)}f. \tag{11}$$

For $\alpha = 30°$ and $v = 0.6\,c$,

$$\cos \alpha = \frac{1}{2}\sqrt{3}\,,\ 1-\frac{v^2}{c^2} = 0.64,\ 1+\frac{v^2}{c^2} = 1.36, \tag{12}$$

so that

$$\frac{f'}{f} = \frac{1.36-0.6\sqrt{3}}{0.64} = 0.5. \tag{13}$$

Thus, there is a decrease of frequency by 50% due to reflection by the moving mirror.

3C. Relativistically Moving Mirror Equation

Fig. 6 – 10 shows the positions of the mirror at time t_0 and t. Since the observer is moving to the left, system is moving relatively to the right. Light beam 1 falls on point a at t_0 and is reflected as beam $1'$. Light beam 2 falls on point d at t and is reflected as beam $2'$. Therefore, $\overline{ab}$ is the wave front of the incoming light at time t_0. The atoms at point are disturbed by the incident wave front $\overline{ab}$ and begin to radiate a wavelet. The disturbance due to the wave front $\overline{ab}$ stops at time t when the wavefront strikes point d. As a consequence

$$\overline{ac} = \overline{bd} = c(t-t_0). \tag{14}$$

From this figure we also have $\overline{ed} = \overline{ag}$, and

$$\sin \alpha = \frac{\overline{bd}+\overline{dg}}{\overline{ag}}\,,\ \sin \beta = \frac{\overline{ac}-\overline{af}}{\overline{ag}-\overline{ef}}. \tag{15}$$

Fig. 6 – 11 displays the beam path 1 in more detail. From this figure it is easy to show that

$$\overline{dg} = \overline{ae} = \frac{\overline{ao}}{\cos \alpha} = \frac{v(t - t_0) \sin \phi}{\cos \alpha} \tag{16}$$

and

$$\overline{af} = \frac{\overline{ao}}{\cos \beta} = \frac{v(t - t_0) \sin \phi}{\cos \beta}. \tag{17}$$

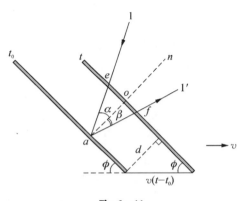

Fig. 6 - 11.

From the triangles aeo and afo we have $\overline{eo} = \overline{ao} \tan \alpha$ and $\overline{of} = \overline{ao} \tan \beta$. Since $\overline{ef} = \overline{eo} + \overline{of}$, then

$$\overline{ef} = v(t - t_0) \sin \phi (\tan \alpha + \tan \beta). \tag{18}$$

By substituting Eq. (14), (16), (17), and (18) into Eq. (15) we obtain

$$\sin \alpha = \frac{c + v \dfrac{\sin \phi}{\cos \alpha}}{\dfrac{\overline{ag}}{t - t_0}}, \tag{19}$$

and

$$\sin \beta = \frac{c - v \dfrac{\sin \phi}{\cos \beta}}{\dfrac{\overline{ag}}{t - t_0} - v \sin \phi \, (\tan \alpha + \tan \beta)}. \tag{20}$$

Eliminating $\dfrac{\overline{ag}}{t - t_0}$ from the two equations above leads to

$$v \sin \phi (\tan \alpha + \tan \beta)$$

$$= c \left(\frac{1}{\sin \alpha} - \frac{1}{\sin \beta} \right) + v\sin \phi \left(\frac{1}{\sin \alpha \cos \alpha} + \frac{1}{\sin \beta \cos \beta} \right). \tag{21}$$

By collecting the terms containing $v \sin \phi$ we obtain

$$\frac{v}{c} \sin \phi \left(\frac{\cos \alpha}{\sin \alpha} + \frac{\cos \beta}{\sin \beta} \right) = \frac{\sin \alpha - \sin \beta}{\sin \alpha \sin \beta} \tag{22}$$

or

$$\sin \alpha - \sin \beta = \frac{v}{c} \sin \phi \, \sin(\alpha + \beta). \tag{23}$$

Experimental Competition

April 28, 2005   Time available: 5 hours

Problem 1
Determination of Shapes by Reflection

Introduction

Direct visual observation, is a method where human beings used their eyes to identify an object. However, not all things in life can be observed directly. For example, how can you tell the position of a broken bone? Is it possible to look at a baby inside a pregnant woman? How about identifying cancer cells inside a brain? All of these require a special technique involving indirect observation.

In this experiment, you are to determine the shape of an object using indirect observation. You will be given two closed cylindrical boxes and in each box, there will be an object with unknown shapes. Your challenge is to reveal the object without opening the box. The physics concepts for this experiment are simple, but creativity and some skills are needed to solve it.

Experiment Apparatus

For this experiment, you will be given two sets of cylindrical boxes consisting of:

(1) An object with unknown shape to be determined (it is a simple geometrical object with either plane or cylindrical sides).

(2) Closed cylindrical box with an angular scale on the top side (2a) and around its circumference (2b).

(3) A knob which you can rotate.

(4) A laser pointer.

(5) Spare batteries for the laser pointer.

(6) A ruler.

Experimental Method

The students are to determine the shape of the object inside a closed cylindrical box. The diameter of the cylinder can be measured by a ruler. Students are not allowed to open the cylindrical box or break the seal to determine the shape of the object. The object is an 8 mm thick metal with its sides polished so that it can reflect light likes a mirror. You can rotate the object using the knob on the top part of the cylinder. This will rotate the object in the same axis as the cylinders axis.

The laser pointer can be switched on by rotating its position. You can adjust the position of the light beam by rotating the laser pointer in either clockwise or anti-clockwise direction. The reflection of the laser beam from the laser pointer can be observed along the circumference of the closed cylinder. Measurement using the angular scale can be used. By rotating the knob on the upper part while the laser pointer is switched on, you will notice that as you rotate the object, the position of the reflected light from the object will change. If the light from the laser pointer dim or the laser pointer fail to work, ask the committee for replacement. By observing the correlation of the angular position of the object and the reflection of the laser beam, you should be able to determine the shape of the object.

For every object (the two objects are of different shapes):

(A) Draw a graph of: reflection angle of the laser beam against the angular position of the object.

(B) Determine the number of edges (sides) in each object.

(C) Use data from the graphic to sketch the shape of the object and find the inside angles positions.

(D) Draw rotating axis of the object and determine the distance to every sides.

(E) Determine the length of sides without error analysis; determine also the angles between neighboring sides.

You must present your result on graph papers and try to deduce the mathematical equations to determine the shape of the object.

Fig. 6 – 12.

Remarks:

(1) One of the objects has only plane sides and the second object has one curved side.

(2) Sometimes you may get two reflections of the beam from the object.

(3) In case of a curved side the determination of the radius of curvature is not required but determination whether it is convex or concave with respect to the axis of rotation is necessary.

Solution

Object Position ($\alpha°$)	Reflected Ray ($\beta°$)	Cal. Reflected Ray ($\beta°$)
-10	211	-149
-5	223	-137
0	230	-130
10	251	-109
20	270	-90
30	290	-70
40	310	-50
50	330	-30
60	349	-11

Cont.

Object Position ($\alpha°$)	Reflected Ray ($\beta°$)	Cal. Reflected Ray ($\beta°$)
75	17	17
80	28	28
90	46	46
100	65	65
110	87	87
120	117	117
130	128	128
140	273	-87
150	291	-69
160	310	-50
170	326	-34
180	343	-17
195	10	10
200	12	12
210	34	34
220	249	-111
230	267	-93
240	285	-75
250	301	-59
260	319	-41
270	337	-23
275	345	-15
290	13	13
300	29	29
310	46	46
320	64	64
330	83	83
340	102	102

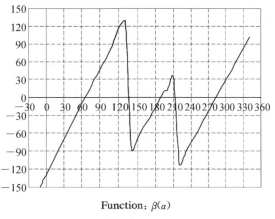

Function: $\beta(\alpha)$

Fig. 6 - 13.

There are 3 jumps on the graph. This is observed at $\alpha = -10°$, $140°$ and $220°$. The jump in the reflection angles are caused by the change of sides, therefore the object has 3 sides and if all the sides are straight sides, we can approximate the lines on the graph using linear regression, i. e.

$$\beta = m\alpha + c,$$

where $\alpha =$ position angle of the object (in °)

and $\beta =$ reflected ray angle (in °)

Segment 1 (-10 to 130): $\beta = 1.98\alpha + c_1,$ (A1)

Segment 2 (140 to 210): $\beta = 1.73\alpha + c_2,$ (A2)

Segment 3 (220 to 340): $\beta = 1.78\alpha + c_3.$ (A3)

To find the gradient, m, as function of side distance from the rotation axis,

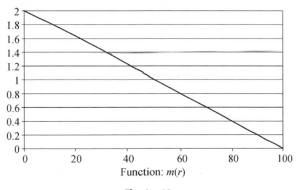

Function: $m(r)$

Fig. 6 - 14.

r, we can simulate it and get a graph and for "small" r:

$$m = -0.02\ r + 2 \text{ or } r = 100 - 50 \text{ m.} \qquad (A4)$$

From (A1) to (A3) and using (A4) we can determine r from the 3 sides:

$$r_1 = 100 - 50(1.98) = 1.5 \text{ mm,}$$
$$r_2 = 100 - 50(1.73) = 13.5 \text{ mm,}$$
$$r_3 = 100 - 50(1.78) = 11.0 \text{ mm.}$$

For each side, we can use the object position when the reflection angle is 0° to draw with a higher precision. The angle for each segment is:

$$\alpha_1 = 66°,$$
$$\alpha_2 = 189°,$$
$$\alpha_3 = 282°.$$

From data obtain the shape of the object can be determined as:

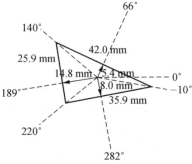

Fig. 6 – 15.

Object Position ($\alpha°$)	Reflected Ray ($\beta°$)	Cal Reflected Ray ($\beta°$)
10	278	− 82
20	296	− 64
30	313	− 47
40	330	− 30
50	347	− 13
65	12	12
70	20	20
80	36	36
90	56	56
100	73	73
110	91	91
120	300	− 60
130	320	− 40

Cont.

Object Position ($\alpha°$)	Reflected Ray ($\beta°$)	Cal Reflected Ray ($\beta°$)
140	342	-18
145	351	-9
155	13	13
160	23	23
170	43	43
180	67	67
190	277	-83
200	297	-63
210	313	-47
220	330	-30
230	347	-13
245	13	13
250	22	22
260	39	39
270	55	55
280	74	74
290	91	91
300	335	-25
305	335	-25
310	336	-24
315	337	-23
320	338	-22
345	21	21
350	22	22
355	22	22
360	23	23
365	23	23

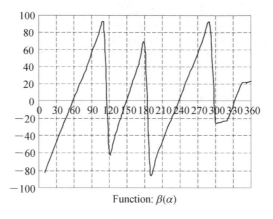

Function: $\beta(\alpha)$

Fig. 6 - 16.

There are 5 jumps in the graphics. This can be observed at $\alpha = 10°$, $120°$, $190°$, $300°$ and $345°$. The jumps in reflection angle are caused by the change of sides, therefore there are 5 sides in the object and if all the sides are straight, we can approximate the lines using linear regression, i. e.

$$\beta = m\alpha + c,$$

where α = angular position of the object (in °)

and β = reflected ray angle (in °)

Segment 1 (10 to 110): $\beta = 1.56\alpha + c_1$,   (B1)

Segment 2 (120 to 180): $\beta = 2.12\alpha + c_2$,   (B2)

Segment 3 (190 to 290): $\beta = 1.64\alpha + c_3$,   (B3)

Segment 4 (190 to 290): $\beta = 0.15\alpha + c_3$,   (B4)

Segment 5 (190 to 290): $\beta = 0.10\alpha + c_3$.   (B5)

From (B1) to (B3) and (A4) we can determine r from the 5 sides:

$$r_1 = 100 - 50(1.56) = 22.0 \text{ mm},$$
$$r_2 = 100 - 50(2.12) = -6.0 \text{ mm},$$
$$r_3 = 100 - 50(1.64) = 18.0 \text{ mm},$$
$$r_4 = 100 - 50(0.15) = 92.5 \text{ mm},$$
$$r_5 = 100 - 50(0.10) = 95.0 \text{ mm}.$$

There are weird data for r_2, r_4 and r_5. It is impossible to have r with either negative or very large value but not so small angle of reflection. So we can guess that it is either a curve sides or double reflection. For double reflection we need to have two adjacent sides with concave angle, so only r_4 and r_5 are possible. So r_2 can only be a curve side. From segment 2 of the graph we can see that the graph looks like a reverse "S" shape, so it is only possible when the sides is concave.

Considering error in the experiment, we can guess that the shape has reflection symmetry.

For each side, we can use the object position when the reflection angle is $0°$ to draw with a higher precision. The angle for segment 1 to 3 is

$$\alpha_1 = 58°,$$
$$\alpha_2 = 149°,$$
$$\alpha_3 = 237°.$$

From the data obtained, the shape of the object can be determined as

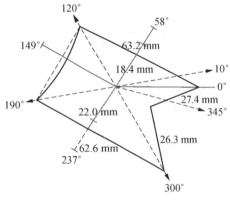

Fig. 6 – 17.

Problem 2
Magnetic Braking on An Inclined Plane

Introduction

When a magnet moves near a non-magnetic conductor such as copper and aluminum, it experiences a dissipative force called magnetic braking force. In this experiment we will investigate the nature of this force.

The magnetic braking force depends on:

— the strength of the magnet, determined by its magnetic moment (μ);

— the conductivity of the conductor (σ_C);

— the size and geometry of both magnet and the conductor;

— the distance between the magnet and conducting surface (d); and

— the velocity of the magnet (v) relative to the conductor.

In this experiment we will investigate the magnetic braking force dependencies on the velocity (v) and the conductor-magnet distance (d). This force can be written empirically as:

$$F_{MB} = -k_0 d^p v^n,\tag{1}$$

where

k_0 is an arbitrary constant that depends on μ, σ_C and geometry of the conductor and magnet which is fixed in this experiment.

d is the distance between the center of magnet to the conductor surface.

v is the velocity of the magnet.

p and n are the power factors to be determined in this experiment.

Experiment

In this experiment error analysis is required.

Apparatus

(1) Doughnut-shaped Neodymium Iron Boron magnet.

     Thickness: $t_M = (6.3 \pm 0.1)$mm

     Outer diameter: $d_M = (25.4 \pm 0.1)$mm

     The poles are on the flat faces as shown:

(2) Aluminum bar (2 pieces).

(3) Acrylic plate for the inclined plane with a linear track for the magnet to roll.

(4) Plastic stand.

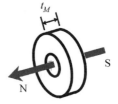

Magnetic poles

Fig. 6 – 18.

(5) Digital stop watch.

(6) Ruler.

(7) Graphic papers (10 pieces).

Additional information:

Local gravitational acceleration: $g = 9.8 \text{ m/s}^2$.

Mass of the magnet: $m = (21.5 \pm 0.5) \text{gram}$.

North-South direction is indicated on the table.

You can read the operation manual of the stopwatch.

This problem is divided into two sections:

(A) Setup and introduction.

(B) Investigation of the magnetic braking force.

Questions

Please provide sufficient diagrams in your answers so that your work can be understood clearly.

(A) Setup

Roll down the magnet along the track as shown. Choose a reasonably small inclination angle so that it does not roll too fast.

(1) As the magnet is very strong, it may experience significant torque due to interaction with earth's magnetic field. It will twist the magnet as it rolls down and may cause significant friction with the track. What will you do to minimize this torque? Explain it using diagram(s).

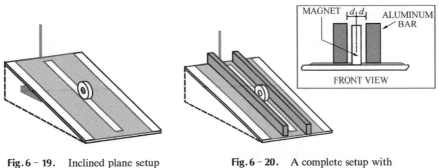

Fig. 6 − 19.   Inclined plane setup without aluminum bars.

Fig. 6 − 20.   A complete setup with aluminum bars.

Place the two aluminum bars as shown in Fig. 6 − 20 with distance approximately $d = 5$ mm. Remember that the distance d is to the center of the magnet as shown in the inset of Fig. 6 - 20.

Again release the magnet and let it roll. You should observe that the magnet would roll down much slower compared to the previous observation due to magnetic braking force.

(2) Provide diagram(s) of field lines and forces to explain the mechanism of magnetic braking.

(B) Investigation of the magnetic braking force

The experimental setup remains the same as shown in Fig. 6 - 20. with the same magnet-conductor distance approximately $d = 5$ mm (about 2 mm gap between magnet and conductor on each side).

(1) Keeping the distance d fixed, investigate the dependence of magnetic braking force on velocity (v). Determine the exponent n of the speed dependence factor in Equation (1). Provide appropriate graph to explain your result.

Now vary the conductor-magnet distance (d) on both left and right. Choose a fixed and reasonably small inclination angle.

(2) Investigate the dependence of the magnetic braking force on conductor-magnet distance (d). Determine the exponent p of the distance dependence factor in Equation (1). Provide appropriate graph to explain your result.

⌖ Solution

(A) Setup and Introduction

(1) To minimize the torque due to interaction of the magnet and the earth's magnetic field we have to set the orientation of the inclined plane so that the magnet will roll down with the poles aligned to the North-South direction as shown.

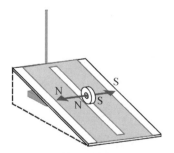

Fig. 6 - 21. Adjusting the orientation of the inclined plane.

(2)

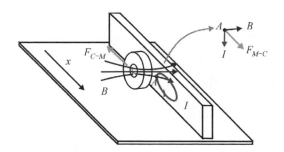

Fig. 6 – 22.   Field and interactions in the magnetic braking effect.

Answer with some vector analysis:

Consider a point A on the conductor. As the magnet moves, its magnetic field sweeps the conductor inducing electric field and causing current flow due to Faraday's law, whose direction can be determined using Lenz's law. Let us choose an arbitrary loop as shown. At point A, the magnetic field and the current will cause Lorentz force F_{M-C} pointing at $x+$ direction. This force is acting on the electrons in the conductor.

On the other hand, due to Newtons' Third law there is reaction force F_{C-M} with the same magnitude but with opposite direction acting on the magnet, which is the magnetic braking force.

(B) Investigation of the magnetic braking force

(1) Determination of the power factor n: Dependence of the magnetic braking force with the velocity.

In this experiment the student has to be aware that the magnet should reach the terminal velocity first before start the timing. From observation we can see that the magnet reaches terminal velocity almost immediately. To make sure we let the magnet travels first for about 5 cm before we start measuring the time. Here we use $s = 250$ mm from start to finish to obtain speed: $v = \dfrac{s}{t}$.

The angle of inclination is varied to take several data. Given $l = 425$ mm, we measure h where $\sin \theta = \dfrac{h}{l}$.

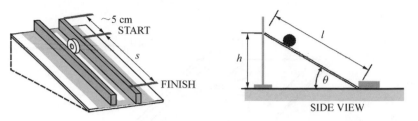

Fig. 6 – 23.   (a) Measurement of the velocity. (b) Measurement of the plane inclination.

Because the magnet-conductor distance is kept constant ($d \approx 5$ mm), the magnetic braking force only depends on the velocity of the magnet, so we can simplify

$$F_{MB} = -k_0 d^p v^n = -k_1 v^n,$$

where $k_1 = k_0 d^p$ is constant in this experiment.

When the magnet reaches the terminal velocity then the total torque should be zero. The equation of the motion at the contact point C will be:

$$\sum \tau_C = 0,$$

$$mg \sin \theta R + F_{MB} R = 0,$$

$$mg \sin \theta - k_1 v^n = 0,$$

$$\sin \theta = \frac{k_1}{mg} v^n.$$

To calculate the power factor n:

$$\ln \sin \theta = \ln\left(\frac{k_1}{mg}\right) + n\ln(v).$$

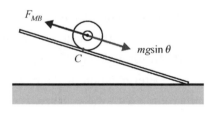

Fig. 6 – 24.   Force diagram of the rolling magnet.

The experimental data:

Table 1. **Experimental data for power factor n determination.**

H(mm)	$\bar{t}$(s)	$\sin \theta$	v(mm/s)	$\ln(v)$	$\ln(\sin \theta)$
23 ± 0.5	22.98 ± 0.005	0.054	10.88	2.39	-2.92
40	12.78	0.094	19.56	2.97	-2.36
50	10.17	0.118	24.58	3.20	-2.14
60	8.62	0.141	29.00	3.37	-1.96
70	6.96	0.165	35.92	3.58	-1.80

Cont.

H(mm)	$\bar{t}$(s)	$\sin\theta$	v(mm/s)	$\ln(v)$	$\ln(\sin\theta)$
80	6. 09	0. 188	41. 05	3. 71	-1.67
91	5. 48	0. 214	45. 62	3. 82	-1.54
101	5. 05	0. 238	49. 50	3. 90	-1.44
111	4. 57	0. 261	54. 70	4. 00	-1.34
120	4. 17	0. 282	59. 95	4. 09	-1.26
130	3. 72	0. 306	67. 20	4. 21	-1.18
150	3. 25	0. 353	76. 92	4. 34	-1.04
170	2. 81	0. 400	88. 97	4. 49	-0.92

Note:

• Column in bold are the data directly taken from the experiment.

• Typical error for h measurement is shown in the first row: $h = (23 \pm 5)$mm. Similar error applies for the rest of h data.

• Data $\bar{t}$ are the average data taken from 3 to 5 measurement. Even though standard deviation error is quite small (± 0.1 s), the error should be dominated by response delay of the observer in pressing the stopwatch. Widely accepted value for human eye response is 0. 25 s, in this experiment we choose more conservative value (± 0.5 s).

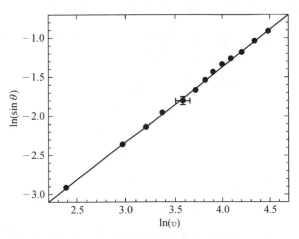

Fig. 6 – 25. Graph of $\ln(\sin\theta)$ vs $\ln(v)$. Typical error bar is shown in the central data.

Using linear regression method or graphical method as shown in Fig. 6 – 22 one can determine n from the slope

$$n = 0.96,$$

whose result is very close to the theoretical value of $n = 1$. From the data shown in Fig. 6 – 16 (as well as the coefficient of correlation $r = 0.9995$), it can be shown that this experiment is very good in demonstrating the linear velocity dependence of the magnetic braking force. This result has been repeated and verified by three independent persons and apparatus setups.

Error estimate of n

Instead of laboring on detailed error propagation analysis that could be very time consuming, in Olympiad context one can make the error estimate as follows:

The typical error of the data points in Fig. 6 – 22 can be obtained from the central data:

$$\ln v = 3.58 \pm 0.075,$$

$$\ln(\sin \theta) = -1.8 \pm 0.075,$$

whose errors propagated from the uncertainties in h and t.

The power factor n can be obtained from the slope of Fig. 6 – 22:

$n = \dfrac{\Delta y}{\Delta x}$ where $y = \ln(\sin \theta)$ and $x = \ln v$.

From the data in Fig. 6 – 22, we have: $\Delta x = 2.1$ and $\Delta y = 2.0$, and the typical errors: $\delta x = 0.075$ and $\delta y = 0.075$.

So the error estimate for n:

$$\frac{\Delta n}{n} = \sqrt{\left(\frac{\delta x}{\Delta x}\right)^2 + \left(\frac{\delta y}{\Delta y}\right)^2} = \sqrt{\left(\frac{0.075}{2.1}\right)^2 + \left(\frac{0.075}{2.0}\right)^2} = 0.05,$$

$$\Delta n = 0.05n = 0.048.$$

So we can conclude the result of our experiment is

$$n = 0.96 \pm 0.05.$$

(2) Determination of the power factor p: Dependence of the magnetic braking force with the magnet-conductor distance.

In this experiment we use one value of inclination angle, $h = 50$ mm ($l = 425$ mm) so that $\theta = \arcsin\left(\dfrac{h}{l}\right) = 6.8°$. Distance travelled remains: $s = 250$ mm, and the timing is done after the magnet travel first for about 5 cm as before.

The equation of motion, similar to previous section:

$$\sum \tau_C = 0,$$

$$mg \sin \theta R + F_{MB}R = 0,$$

$$mg \sin \theta - k_0 d^p v^n = 0,$$

$$v^{-n} = \frac{k_0}{mg \sin \theta} d^p,$$

$$-n\ln v = \ln\left(\frac{k_0}{mg \sin \theta}\right) + p\ln d.$$

Here again p can be obtained using linear regression or graphical method where we use the previously obtained value: $n = 0.96 \pm 0.05$.

The experimental data:

Table 2. Experimental data for power factor p determination.

d(mm)	t(s)	v(mm/s)	$-n\ln v$	$\ln d$
4.5 ± 0.5	13.53 ± 0.005	18.48	-2.80	1.50
5.5	9.60	26.04	-3.13	1.70
6.5	6.70	37.31	-3.47	1.87
7.5	4.99	50.10	-3.76	2.01
8.5	3.47	72.05	-4.11	2.14
9.5	2.87	87.11	-4.29	2.25
10.5	2.14	116.82	-4.57	2.35
11.5	1.66	150.60	-4.81	2.44

Note:

Distance d is measured from the center of the magnet.

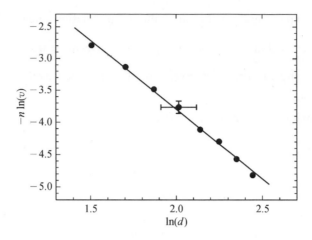

Fig. 6 – 26. Graph of $-n\ln(v)$ vs $\ln(d)$. Typical error
bar is shown in the central data.

From linear regression calculation we have

$$p = -2.16.$$

So the magnetic braking force is very sensitive with the magnet — conductor
distance d in which the relationship is almost inversely quadratic. In brief,
the further the magnet from the conductor the weaker the magnetic braking
force becomes. This result has been repeated and verified by three
independent persons and apparatus setups.

Error estimate of p

Similar to previous section, we use the central data shown in Fig. 6 – 24.

$$\ln d = 2.01 \pm 0.105,$$
$$-n\ln(v) = -3.76 \pm 0.095.$$

The power factor p can be obtained from the slope of line in di Fig. 6 – 24.

$p = \dfrac{\Delta y}{\Delta x}$ where $y = -n\ln v$ and $x = \ln d$.

For the data shown in Fig. 6 – 24, we obtain: $\Delta x = 0.94$ dan $\Delta y = 2.01$,
with typical error: $\delta x = 0.105$ and $\delta y = 0.095$.
So the error estimate for p:

$$\frac{\Delta p}{p} = \sqrt{\left(\frac{\delta x}{\Delta x}\right)^2 + \left(\frac{\delta y}{\Delta y}\right)^2} = \sqrt{\left(\frac{0.105}{0.94}\right)^2 + \left(\frac{0.095}{2.01}\right)^2} = 0.12,$$

$$\Delta p = 0.12p = 0.26.$$

So we can conclude the result of our experiment is

$$p = -2.2 \pm 0.3.$$

Minutes of the Seventh Asian Physics Olympiad

Almaty (Kazakhstan), April 22 – 30, 2006

1. The following 18 countries were present at the 7th Asian Physics Olympiad:

Australia (8+2), Azerbaijan (5+2), Vietnam (8+2), Georgia (3+1), Israel (8+2), India (team A: 8+2; team B: 2+1), Indonesia (team A: 8+2, team B: 8+2), Jordan (5+2), Kazakhstan (team A: 8+2, team B: 8+2), Cambodia (4+2), China (8 + 2), Kyrgyzstan (7 + 2), Mongolia (6 + 2), Singapore (8+2), Tajikistan (5+1), Thailand (8+2), Chinese Taipei (8+2), Turkmenistan (3+2). The first number in parentheses denotes the number of the competitors while the second number denotes the number of the leaders.

2. Results of marking the papers by the Organizers were presented:

The best score was 44.3 points achieved by Zhu Li from China (Absolute winner of the 7th APhO). The second best score (43.8 points) was achieved by Wang Xingze from China. The third best result (42.7 points) was reached *ex aequo* by Yang Shuolong (China) and Ho Pangus (Indonesia). The following limits for awarding the medals and honorable mentions were established according to the Statutes:

Gold Medal:	39 points,
Silver Medal:	34 points,
Bronze Medal:	28 points,
Honorable Mention:	21 points.

According to the above limits 12 Gold Medals, 11 Silver Medals, 25 Bronze Medals and 27 Honorable Mentions were awarded.

The results of the 7th APhO have been distributed to all the delegations.

3. In addition to the regular prizes, the following special prizes were awarded:

● for the Absolute Winner: Zhu Li, China

● for the best score in the theoretical part of the competition: Wang Xingze, China

● for the best score in the experimental part of the competition: Ho Pangus, Indonesia

● for the best score among the female participants: Xu Lilei, China

● for the youngest participant: Karthik Ganapathy, India

- for the best score among the participants from the teams that joined the APhO this year (President's Prize): Rai Ankur, India
- for the highest determination to win: George Mikoshevich, Georgia
- for the 11 Kazakh participants from teams A and B who finish high school this year (scholarships founded by the Kazakh National Technical University):

> Levchenko Pavel
>
> Agybay Kuanysh
>
> Kaipov Yermek
>
> Bychkov Vladimir
>
> Urozayev Dias
>
> Saurbayev Ilias
>
> Shagyrov Olzhas
>
> Anenkov Vasiliy
>
> Pak Michail
>
> Faizullin Ruzel
>
> Murzagaliyev Yerlan

4. The International Board unanimously reelected Prof. Ming-Juey Lin to the post of Secretary of the Asian Physics Olympiads for the next five years.

5. President and Secretary of the APhOs presented the current list of countries to host the competitions in the future:

- 2007 – China (Shanghai) (invitations made)
- 2008 – Mongolia (proposed)
- 2009 – Thailand (preliminary contact)
- 2010 – Chinese Taipei (confirmed)
- 2011 – Israel (confirmed)
- 2012 – Cambodia (preliminary contact)

6. President of the APhOs suggested taking five first results (instead of three ones) in defining 100% in the awarding system. After preliminary discussion there has been decided that a final wording of the proposal will be disseminated by the Secretariat of the APhOs to all the participating countries according to the Statutes. It was suggested to vote the proposal at the first session of the International Board at the Olympiad in China.

7. The International Board expressed deep thanks to the Authorities of the

Republic of Kazakhstan, Administration of Almaty city and Almaty oblast, Center Daryn and Mrs. Tildash Bituova and her collaborators for excellent preparing and executing the competition. The International Board highlighted the high level of the competition and the very friendly atmosphere it was accompanied with.

8. Acting on behalf of the Organizers of the next Asian Physics Olympiad, Professor Zuimin Jiang announced that the 8th APhO would be held in Shanghai (China) on April 21 – 29, 2007 and cordially invited the participating countries to attend the competition.

Almaty, April 30, 2006.

Waldemar Gorzkowski　　　　　　**Ming-Juey Lin, Ph. D.**
Honorary President　　　　　　Secretary of the APhOs
of the APhOs

Mrs. Tildash Bituova　　　　　　**Yohanes Surya, Ph. D.**
Vice-President　　　　　　President of the APhOs
of the Organizing Committee
of the 7th APhO

Theoretical Competition

April 24, 2006   Time available: 5 hours

Problem 1

This problem consists of four not related parts.

(A) Two identical conducting plates α and β with charges $-Q$ and $+q$ respectively $(Q > q > 0)$ are located parallel to each other at a small distance. Another identical plate γ with mass m and charge $+Q$ is situated parallel to the original plates at distance d from the plate β (see Fig. 7 - 1).

Surface area of the plates is S. The plate γ is released from rest and can move freely, while the plates α and β are kept fixed. Assume that the collision between the plates β and γ is elastic, and neglect the gravitational force and the boundary effects. Assume that the charge has enough time to redistribute between plates β and γ during the collision.

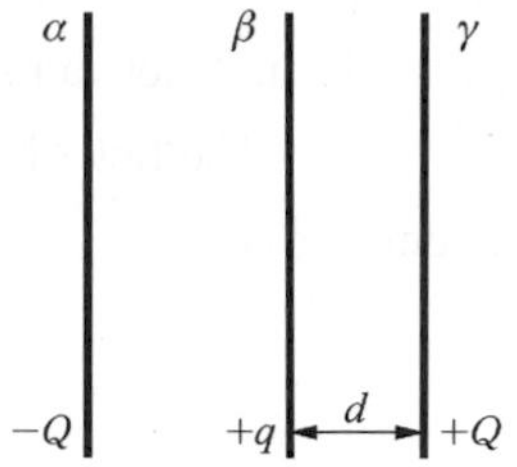

Fig. 7 - 1.

(A1) What is the electric field E_1 acting on the plate γ before the collision with the plate β?

(A2) What are the charges of the plates Q_β and Q_γ after the collision?

(A3) What is the velocity v of the plate γ after the collision at the distance d from the plate β?

(B) Massless mobile piston separates the vessel into two parts. The vessel is isolated from the environment. One part of the vessel contains $m_1 = 3.00$ g of diatomic hydrogen at the temperature of $T_{10} = 300$ K, and the other part contains $m_2 = 16.00$ g of diatomic oxygen at the temperature of $T_{20} = 400$ K. Molar masses of hydrogen and oxygen are $\mu_1 = 2.00$ g/mole and $\mu_2 = 32.00$ g/mole respectively, and $R = 8.31$ J/(K · mole). The piston weakly conducts heat between oxygen and hydrogen, and eventually the temperature in the system equilibrates. All the processes are quasi stationary.

(B1) What is the final temperature of the system T?

(B2) What is the ratio between final pressure P_f and initial pressure P_i?

(B3) What is the total amount of heat Q, transferred from oxygen to hydrogen?

(C) The Mariana Abyss in the Pacific Ocean has a depth of $H = 10\,920$ m. Density of salted water at the surface of the ocean is $\rho_0 = 1025$ kg/m^3, bulk modulus is $K = 2.1 \cdot 10^9$ Pa, acceleration of gravity is $g = 9.81$ m/s^2. Neglect the change in the temperature and in the acceleration of gravity with the depth, and also neglect the atmospheric pressure. Find the numerical value of the pressure $P(H)$ at the bottom of the Mariana Abyss. You may use exact or iterative methods. In the latter case you may keep only the first nonvanishing term in compressibility.

Note: The fluids have very small compressibility. Compressibility coefficient is defined as

$$\alpha = -\frac{1}{V}\left(\frac{dV}{dp}\right)_{T=\text{const}}.$$

Bulk modulus K is the inverse of α: $K = \dfrac{1}{\alpha}$.

(D) Two thin lenses with lens powers D_1 and D_2 are located at distance $L = 25$ cm from each other, and their main optical axes coincide. Lens power is the inverse of focal length. This system creates a direct real image of the object, located at the main optical axis closer to lens D_1, with the magnification $\Gamma' = 1$. If the positions of the two lenses are exchanged, the system again produces a direct real image, with the magnification $\Gamma'' = 4$.

(D1) What are the types of the lenses? On the answer sheet you should mark the gathering lens as « $+$ », and the diverging lens as « $-$ ». Include diagrams to illustrate your answers.

(D2) What is the difference between the lens powers $\Delta D = D_1 - D_2$?

🔑 Solution

(A) The electric field acting on the plate γ before the collision is

(A1) $$E_1 = \frac{Q-q}{2\varepsilon_0 S}.$$ (1)

The force acting on the plate is

$$F_1 = E_1 Q = \frac{(Q-q)Q}{2\varepsilon_0 S}.$$ (2)

The work done by the electric field before the collision is

$$A_1 = F_1 d = \frac{(Q-q)Qd}{2\varepsilon_0 S}.$$ (3)

The charge will get redistributed between two touching conducting plates during the collision. The values of the charges can be obtained from the condition that the electric field between the touching plates vanishes. If one assumes that the plate γ is on the right side, the left surface of the combined plate will have the charge

(A2a) $$Q_\beta = Q + \frac{q}{2},$$ (4)

and the right surface will have the charge

(A2b) $$Q_\gamma = \frac{q}{2}.$$ (5)

These charges remain on the plates after the collision is over. Now the force acting on the plate γ equals $F_2 = \frac{E_2 q}{2}$, where $E_2 = \frac{\frac{q}{2}}{2\varepsilon_0 S}$. The work done by field E_2 is

$$A_2 = F_2 d = \frac{q^2 d}{8\varepsilon_0 S}.$$ (6)

The total work done by the electric fields is

$$A = A_1 + A_2 = \frac{d}{2\varepsilon_0 S}\left(Q - \frac{q}{2}\right)^2.$$ (7)

Velocity at the distance d can be calculated using the following relation:

$$\frac{mv^2}{2} = A.$$ (8)

Substituting (8) into (7), we finally get

(A3)
$$v = \left(Q - \frac{q}{2}\right)\sqrt{\frac{d}{m\varepsilon_0 S}}.$$
(9)

(**B**) The total work done by the gases is zero. Thus at any moment the total internal energy equals the original value:

$$\frac{m_1}{\mu_1}C_V T_1 + \frac{m_2}{\mu_2}C_V T_2 = \frac{m_1}{\mu_1}C_V T_{10} + \frac{m_2}{\mu_2}C_V T_{20},$$
(10)

where $\mu_1 = 2$ g/mole and $\mu_2 = 32$ g/mole are molar masses of hydrogen and oxygen, and $C_V = \frac{5R}{2}$ is the molar heat capacity of diatomic gas. The final temperature of the system is

(B1)
$$T = \frac{\frac{m_1}{\mu_1}T_{10} + \frac{m_2}{\mu_2}T_{20}}{\frac{m_1}{\mu_1} + \frac{m_2}{\mu_2}} = 325 \text{ K.}$$
(11)

The temperature of oxygen decreases, and the amount of heat Q is transferred to hydrogen by heat conduction. The piston will move in the direction of the oxygen, thus the hydrogen does a positive work $A > 0$, and the change of the internal energy of oxygen is $\Delta U = A - Q$. On the other hand,

$$\Delta U = \frac{m_2}{\mu_2}\frac{5}{2}R(T - T_{20}) = -779 \text{ J.}$$
(12)

To find A, let us prove that the pressure p does not change. Differentiating the equations of the state for each gas, we get

$$\Delta T_1 = \frac{\mu_1}{m_1 R}(p\Delta V + V_1\Delta p), \quad \Delta T_2 = \frac{\mu_2}{m_2 R}(-p\Delta V + V_2\Delta p),$$
(13)

where V_i are the gas volumes, and $\Delta V = \Delta V_1 = -\Delta V_2$ is the change of the volume of the hydrogen. Differentiating (10), we get

$$\frac{m_1}{\mu_1}\Delta T_1 + \frac{m_2}{\mu_2}\Delta T_2 = 0.$$
(14)

Substituting (13) into (14), we obtain $(V_1 + V_2) \cdot \Delta p = 0$, thus

(B2)
$$\frac{p_f}{p_i} = 1. \tag{15}$$

Then the work done by the hydrogen is

$$A = p \cdot \Delta V = -\frac{m_2}{\mu_2} R \cdot \Delta T_2 = \frac{m_2}{\mu_2} R(T_{20} - T) = 312 \text{ J}. \tag{16}$$

The total amount of heat transferred to hydrogen is

(B3)
$$Q = A - \Delta U = 1091 \text{ J}. \tag{17}$$

(C) The change of the pressure is related to the change in the density via

$$\Delta p = -K \frac{\Delta V}{V} = K \frac{\Delta \rho}{\rho} \approx K \frac{\Delta \rho}{\rho_0}, \tag{18}$$

where ρ_0 is the density of water at the surface.

$$\rho = \rho_0 + \Delta \rho = \rho_0 \left(1 + \frac{\Delta \rho}{\rho_0}\right) = \rho_0 \left(1 + \frac{\Delta p}{K}\right), \tag{19}$$

where $\Delta p \approx p$ (we neglect the atmospheric pressure). Then

(C1)
$$\rho(x) = \rho_0 \left[1 + \frac{p(x)}{K}\right]. \tag{20}$$

The change of the hydrostatic pressure with the depth equals

$$dp = g \cdot \rho(x) dx, \quad \frac{dp}{dx} = g\rho(x) = g\rho_0 + g\rho_0 \frac{p(x)}{K}, \tag{21}$$

$$\frac{dp(x)}{dx} - \frac{g\rho_0}{K} p(x) = g\rho_0. \tag{22}$$

The solution of this differential equation with boundary condition $p(0) = 0$ is

$$p(x) = K \left(\exp \frac{g\rho_0}{K} x - 1\right). \tag{23}$$

Since $\frac{g\rho_0}{K} H \ll 1$, we can use the expansion

$$\exp z \approx 1 + z + \frac{z^2}{2!} + \ldots, \tag{24}$$

thus

$$p(x) \cong g\rho_0 x + \frac{1}{2K}(g\rho_0 x)^2. \tag{25}$$

The last formula can be simply derived using the method of successive iterations. First, the pressure can be estimated without compressibility taken into account:

$$p_0(x) = g\rho_0 x. \tag{26}$$

Correction to the density in the first approximation can be obtained using $p_0(x)$:

$$\rho_1(x) = \rho_0 \left(1 + \frac{g\rho_0 x}{K}\right). \tag{27}$$

Now, correction to pressure can be obtained using $\rho_1(x)$:

$$p_1(H) = \int_0^H \rho_1(x) g \, \mathrm{d}x = g\rho_0 x + \frac{1}{2K}(g\rho_0 x)^2, \tag{28}$$

as obtained earlier.

Putting in the numerical values, we get

(C2) $\qquad p(H) = (1098 \times 10^5 + 28.7 \times 10^5)\,\mathrm{Pa}$

$$\approx 1.13 \times 10^8 \, \mathrm{Pa}. \tag{29}$$

(**D**) First one has to determine the types of the lenses. If both lenses are negative, one always obtains a direct imaginary image. If one lens is positive and the other is negative, three variants are possible: an inversed real image, a direct imaginary image or an inversed imaginary image, all contradicting the conditions of the problem. Only the last variant is left – two positive lenses. The first lens creates an inversed real image, and the second one inverts in once more, creating the direct real image. Using the lens equations, the magnifications of the lenses can be written as

$$\Gamma_1 = \frac{F_1}{d_1 - F_1}; \qquad \Gamma_2 = \frac{F_2}{d_2 - F_2}, \tag{30}$$

where d_1 is the distance from the object to the first lens, $d_2 = L - f_1$ is the distance from the image of the first lens to the second lens, and f_1 is the distance from the first lens to the first image. The total magnification of the system is $\Gamma' = \Gamma_1 \cdot \Gamma_2$. Using the expression for d_2, inverted magnification coefficient can be written as

$$\frac{1}{\Gamma'} = \frac{d_1 [L - (F_1 + F_2)]}{F_1 F_2} - \frac{L}{F_2} + 1. \tag{31}$$

One notices from this expression that if two lenses are exchanged, the first term stays invariant, and only the second term changes. Thus the expression for the inverted magnification in the second case is:

$$\frac{1}{\Gamma''} = \frac{d_1 [L - (F_1 + F_2)]}{F_1 F_2} - \frac{L}{F_1} + 1. \tag{32}$$

Subtracting these two formulas, we get:

$$\frac{1}{\Gamma'} - \frac{1}{\Gamma''} = L\left(\frac{1}{F_1} - \frac{1}{F_2}\right) = L(D_1 - D_2); \tag{33}$$

$$D_1 - D_2 = \frac{1}{L}\left(\frac{1}{\Gamma'} - \frac{1}{\Gamma''}\right) = \frac{1}{0.25}\left(1 - \frac{1}{4}\right)$$

$$= \frac{1}{0.25} \cdot \frac{3}{4} = 3 \text{ diopters.} \tag{34}$$

Problem 2
Oscillator Damped by Sliding Friction

Theoretical Introduction

In mechanics, one often uses so called phase space, an imaginary space with the axes comprising of coordinates and moments (or velocities) of all the material points of the system. Points of the phase space are called imaging points. Every imaging point determines some state of the system.

When the mechanical system evolves, the corresponding imaging point follows a trajectory in the phase space which is called phase trajectory. One puts an arrow along the phase trajectory to show direction of the evolution. A set of all possible phase trajectories of a given mechanical system is called a phase portrait of the system. Analysis of this phase portrait allows one to unravel important qualitative properties of dynamics of the system, without solving equations of motion of the system in an explicit form. In many cases, the use of the phase space is the most appropriate method to solve problems in mechanics.

In this problem, we suggest you to use phase space in analyzing some mechanical systems with one degree of freedom, i. e. systems which are described by only one coordinate. In this case, the phase space is a two-dimensional plane. The phase trajectory is a curve on this plane given by a dependence of the momentum on the coordinate of the point, or vice versa, by a dependence of the coordinate of the point on the momentum.

As an example we present a phase trajectory of a free particle moving along x axis in positive direction (Fig. 7 – 2).

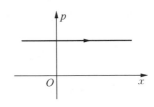

Fig. 7 – 2. Phase trajectory of a free particle.

Questions

(A) Phase portraits

(A1) Make a draw of the phase trajectory of a free material point moving between two parallel absolutely reflective walls located at $x = -\dfrac{L}{2}$ and $x = \dfrac{L}{2}$.

(A2) Investigate the phase trajectory of the harmonic oscillator, i. e. of the material point of mass m affected by Hook's force $F = -kx$:

(a) Find the equation of the phase trajectory and its parameters.

(b) Make a draw of the phase trajectory of the harmonic oscillator.

(A3) Consider a material point of mass m on the end of weightless solid rod of length L, another end of which is fixed (strength of gravitational field is g). It is convenient to use the angle α between the rod and vertical line as a coordinate of the system. The phase plane is the plane with coordinates $\left(\alpha, \dfrac{d\alpha}{dt} \right)$. Study and make a draw of the phase portrait of this pendulum at arbitrary angle α. How many qualitatively different types of phase trajectories K does this system have? Draw at least one typical trajectory of each type. Find the conditions which determine these different types of phase trajectories. (Do not take the equilibrium points as phase trajectories). Neglect air resistance.

(B) The oscillator damped by sliding friction

When considering resistance to a motion, we usually deal with two types of friction forces. The first type is the friction force, which depends on the velocity (viscous friction), and is defined by $F = -\gamma v$. An example is given by a motion of a solid body in gases or liquids. The second type is the friction force, which does not depend on the magnitude of velocity. It is defined by the value $F = \mu N$ and direction opposite to the relative velocity of contacting bodies (sliding friction). An example is given by a motion of a solid body on the surface of another solid body.

As a specific example of the second type, consider a solid body on a horizontal surface at the end of a spring, another end of which is fixed. The mass of the body is m, the elasticity coefficient of the spring is k, the friction coefficient between the body and the surface is μ. Assume that the body moves along the straight line with the coordinate x ($x = 0$ corresponds to the spring which is not stretched). Assume that static and dynamical friction coefficients are the same. At initial moment the body has a position $x = A_0 (A_0 > 0)$ and zero velocity.

(B1) Write down equation of motion of the harmonic oscillator damped by the sliding friction.

(B2) Make a draw of the phase trajectory of this oscillator and find the equilibrium points.

(B3) Does the body completely stop at the position where the string is not stretched? If not, determine the length of the region where the body can come to a complete stop.

(B4) Find the decrease of the maximal deviation of the oscillator in positive x direction during one oscillation ΔA. What is the time between two consequent maximal deviations in positive direction? Find the dependence of this maximal deviation $A(t_n)$ where t_n is the time of the n-th maximal deviation in positive direction.

(B5) Make a draw of the dependence of coordinate on time, $x(t)$, and estimate the number N of oscillations of the body?

Note:

Equation of the ellipse with semi-axes a and b and centre at the origin has the following form:

$$\frac{x^2}{a^2} + \frac{y^2}{b^2} = 1.$$

 Solution

(A) **Phase portraits**

(A1) Let Ox axis be pointed perpendicular to the walls. Since the material point is free and collisions are absolutely elastic then the magnitude of momentum is conserved, while its direction is changed to opposite at the collisions. Hence, the phase trajectory is of the following form (Fig. 7 – 3):

The motion with positive values of the momentum is directed along increasing values of the coordinate. Thus, the phase trajectory is directed clockwise, as indicated in Fig. 7 – 3.

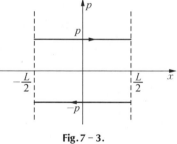

Fig. 7 – 3.

(A2)

(a) For the harmonic oscillator, let us denote the coordinate by x, the momentum by p, and the total energy by E. The energy conservation law is

$$\frac{p^2}{2m} + \frac{kx^2}{2} = E.$$

This expression determines the equation of phase trajectory, for a given E. Dividing both sides of the equation by E, we obtain

$$\frac{p^2}{2mE} + \frac{x^2}{\frac{2E}{k}} = 1.$$

This is a canonical form of the equation of ellipse in (x, p). The centre of the ellipse is at $(0, 0)$ and the semiaxes are $\sqrt{\dfrac{2E}{k}}$ and $\sqrt{2Em}$ respectively.

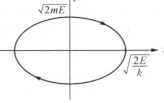

Fig.7 − 4.

(b) The phase trajectory is of the following form in Fig.7 − 4.

The motion with positive values of the momentum is directed along increasing values of the coordinate. Thus, the phase trajectory is directed clockwise, as indicated in Fig.7 − 3.

(A3) Let us choose the potential energy level at the lowest point of the pendulum (equilibrium state). Taking into account for that linear velocity of the point is $v = L\dot{\alpha}$, we write down the total energy of the mathematical pendulum

$$\frac{mL^2\dot{\alpha}^2}{2} + mgL(1 - \cos\alpha) = E.$$

Analysis of this expression leads to the following:

at $E < 2mgL$ the pendulum oscillates about the lower equilibrium position; if $E \ll mgL$, the oscillations are harmonic;

at $E = 2mgL$ the pendulum does not oscillate; the pendulum tends to the upper point of equilibrium.

at $E > 2mgL$ the pendulum rotates about fixed point.

The phase trajectory is shown in Fig.7 − 5.

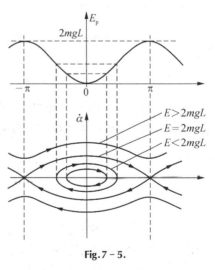

Fig.7 − 5.

There are $K = 3$ qualitatively different types of the phase trajectories: oscillations, rotations, and the motion to the upper point of equilibrium (separatrisse). (We do not take the equilibrium points as phase trajectories.)

(B) The oscillator damped by sliding friction

(B1) For the sliding friction the `magnitude of the friction force does not depend on the magnitude of velocity, but its direction is opposite to the velocity vector of the body. Therefore, the equations of motion should be written separately for the motion to the right and to the left from the "equilibrium" point (the spring is not stretched). Let us choose the x-axis along the direction of motion, and the origin of the coordinate system at the equilibrium point without the friction force. We obtain the equations of motion as follows:

$$\ddot{x} + \omega_0^2 x = -\frac{F_{fr}}{m}, \quad \dot{x} > 0,$$

$$\ddot{x} + \omega_0^2 x = \frac{F_{fr}}{m}, \quad \dot{x} < 0. \tag{1}$$

Here, $F_{fr} = \mu mg$ is the friction force, $\omega_0^2 = \frac{k}{m}$ is the frequency of oscillations of the pendulum without the friction.

(B2) Introducing the variables $x_1 = x + \frac{F_{fr}}{m\omega_0^2}$ and $x_2 = x - \frac{F_{fr}}{m\omega_0^2}$ we can write down the equations of motion in the same form for both the cases,

$$\ddot{x}_{1,2} + \omega_0^2 x_{1,2} = 0,$$

which coincides with the equation of motion of harmonic oscillator without the friction. The action of the friction force is reduced to a drift of the equilibrium points: for $\dot{x} > 0$, it becomes $x_- = -\frac{F_{fr}}{m\omega_0^2}$, $x_1 = 0$ and for $\dot{x} < 0$ it becomes $x_+ = \frac{F_{fr}}{m\omega_0^2}$, $x_2 = 0$.

Thus, due to the section (A2) above, the phase trajectory is a combination of parts of ellipses with centers at the point x_- for an upper half-plane $p > 0$, and at the point x_+ for a lower half-plane $p < 0$. As the result of a continuity of motion these parts of ellipses should comprise a continuous curve by meeting each other at $p = 0$.

Thus, the phase trajectory is Fig. 7 - 6.

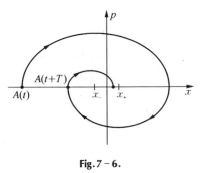

Fig. 7 - 6.

(B3) According to the phase trajectory combination, the body not necessarily stops at the point $x = 0$. It will stop when it falls into the range from x_- to x_+. This region is called stagnation region. The width of this region is

$$x_+ - x_- = \frac{2F_{fr}}{m\omega_0^2}.$$

(B4) From the definition of equilibrium points and the obtained form of phase trajectory, it is easy to find the decrease of amplitude during one period:

$$\Delta A = A(t) - A(t + T) = 2(x_+ - x_-) = \frac{4F_{fr}}{m\omega_0^2}.$$

This can be rewritten as

$$A(t) - A(t + T) = \frac{4F_{fr}}{2\pi m\omega_0} T.$$

One can see that, unlike the case of viscous friction, the amplitude decreases in accord to a linear law, $A = A_0 - pt_n$, where $p = \frac{2F_{fr}}{\pi m\omega_0}$.

(B5) The total number of oscillations depends on the initial amplitude A_0, and it can be found as

$$N = \frac{A_0}{2(x_+ - x_-)}.$$

As the result of the above conclusions the plot of $x(t)$ is of the following form in Fig. 7 - 7.

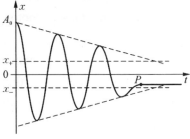

Fig. 7 - 7.

The frequency is equal to the frequency of free oscillator, $\omega_0^2 = \frac{k}{m}$. The

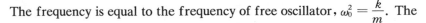

time between two successive maximal deviations is $T = \frac{2\pi}{\omega_0}$.

The oscillations do not stop until the amplitude is more than half-width of the stagnation region $x_+ - x_-$. In real situations, the body stops in random positions within the stagnation region. In Fig. 7 - 7 the point P denotes the point where the body stops.

Problem 3
Laser Cooling of Atoms

In this problem you are asked to consider the mechanism of atom cooling with the help of laser radiation. Investigations in this field led to considerable progress in the understanding of the properties of quantum gases of cold atoms, and were awarded Nobel prizes in 1997 and 2001.

Theoretical Introduction

Consider a simple two-level model of the atom, with ground state energy E_g and excited state energy E_e. Energy difference is $E_e - E_g = \hbar\omega_0$, the angular frequency of used laser is ω, and the laser detuning is $\delta = \omega - \omega_0 \ll \omega_0$. Assume that all atom velocities satisfy $v \ll c$, where c is the light speed. You can always restrict yourself to first nontrivial orders in small parameters $\frac{v}{c}$ and $\frac{\delta}{\omega_0}$. Natural width of the excited state E_e due to spontaneous decay is $\gamma \ll \omega_0$: for an atom in an excited state, the probability to return to a ground state per unit time equals γ. When an atom returns to a ground state, it emits a photon of a frequency close to ω_0 in a random direction.

It can be shown in quantum mechanics, that when an atom is subject to low-intensity laser radiation, the probability to excite the atom per unit time depends on the frequency of radiation in the reference frame of the atom, ω_a, according to

$$\gamma_p = s_0 \frac{\frac{\gamma}{2}}{1 + \frac{4(\omega_a - \omega_0)^2}{\gamma^2}} \ll \gamma,$$

where $s_0 \ll 1$ is a parameter, which depends on the properties of atoms and laser intensity.

In this problem properties of the gas of sodium atoms are investigated neglecting the interactions between the atoms. The laser intensity is small enough, so that the number of atoms in the excited state is always much smaller than number of atoms in the ground state. You can also neglect the effects of the gravitation, which are compensated in real experiments by an additional magnetic field.

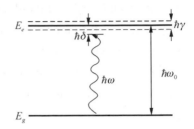

Fig. 7 - 8. Note that shown parameters are not in scale.

Numerical values:

Planck constant	$\hbar = 1.05 \cdot 10^{-34}\,\text{Js}$
Boltzmann constant	$k_B = 1.38 \cdot 10^{-23}\,\text{JK}^{-1}$
Mass of sodium atom	$m = 3.81 \cdot 10^{-26}\,\text{kg}$
Frequency of used transition	$\omega_0 = 2\pi \cdot 5.08 \cdot 10^{14}\,\text{Hz}$
Excited state linewidth	$\gamma = 2\pi \cdot 9.80 \cdot 10^{6}\,\text{Hz}$
Concentration of the atoms	$n = 10^{14}\,\text{cm}^{-3}$

Questions

(a) Suppose the atom is moving in the positive x direction with the velocity v_x, and the laser radiation with frequency ω is propagating in the negative x direction. What is the frequency of radiation in the reference frame of the atom?

(b) Suppose the atom is moving in the positive x direction with the velocity v_x, and two identical laser beams shine along x direction from different sides. Laser frequencies are ω, and intensity parameters are s_0. Find the expression for the average force $F(v_x)$ acting on an atom. For small v_x this force can be written as $F(v_x) = -\beta v_x$. Find the expression for β. What is the sign of $\delta = \omega - \omega_0$, if the absolute value of the velocity of the atom decreases? Assume that momentum of an atom is much larger than the momentum of a photon.

In what follows we will assume that the atom velocity is small enough so that one can use the linear expression for the average force.

(c) If one uses 6 lasers along x, y and z axes in positive and negative directions, then for $\beta > 0$ the dissipative force acts on the atoms, and their average energy decreases. This means that the temperature of the gas, which is defined through the average energy, decreases. Using the concentration of the atoms given above, estimate numerically the temperature T_Q, for which one cannot consider atoms as point-like objects because of quantum effects.

In what follows we will assume that the temperature is much larger than T_Q and six lasers along x, y and z directions are used, as was explained in part (c).

In part (b) you calculated the average force acting on the atom. However, because of the quantum nature of photons, in each absorption or emission process the momentum of the atom changes by some discrete value and in random direction, due to the recoil processes.

(d) Determine numerically the square value of the change of the momentum of the atom, $(\Delta p)^2$, as the result of one absorption or emission event.

(e) Because of the recoil effect, average temperature of the gas after long time does not become an absolute zero, but reaches some finite value. The evolution of the momentum of the atom can be represented as a random walk in the momentum space with an average step $\sqrt{(\Delta p)^2}$, and a cooling due to the dissipative force. The steady-state temperature is determined by the combined effect of these two different processes. Show that the steady state temperature T_d is of the form: $T_d = \dfrac{\hbar\gamma\left(x + \dfrac{1}{x}\right)}{4k_{\mathrm{B}}}$. Determine x.

Assume that T_d is much larger than $\dfrac{(\Delta p)^2}{2k_{\mathrm{B}}m}$.

Note: If vectors P_1, P_2, $\cdots$, P_n are mutually statistically uncorrelated, mean square value of their sum is

$$\langle (P_1 + P_2 + \cdots + P_n)^2 \rangle = \langle P_1{}^2 \rangle + \langle P_2{}^2 \rangle + \ldots + \langle P_n{}^2 \rangle.$$

(f) Find numerically the minimal possible value of the temperature due to recoil effect. For what ratio $\dfrac{\delta}{\gamma}$ is it achieved?

Solution

(a) $\omega\left(1+\dfrac{v_x}{c}\right)$, this is classic Doppler effect.

(b) Absolute value of the momentum, transferred during each absorption, equals

$$\frac{\hbar\omega_0}{c}. \tag{1}$$

The momentum of the emitted photon is uniformly distributed over different directions, and after averaging gives a contribution which is much smaller than $\dfrac{\hbar\omega_0}{c}$. The average force is nonzero, since for atoms moving towards right frequency of right laser gets larger (due to Doppler effect discussed in part (A)), while frequency of left laser goes down. Since number of scattered photons depends on the frequency in the reference frame of the atom, there is a net nonzero force. It equals

$$F(v_x) = F_+ + F_-$$

$$= -\frac{\hbar\omega_0}{c} \cdot \frac{s_0 \gamma}{2} \cdot \left[\frac{1}{1+\dfrac{4\left(\delta+\dfrac{\omega_0 v_x}{c}\right)^2}{\gamma^2}} - \frac{1}{1+\dfrac{4\left(\delta-\dfrac{\omega_0 v_x}{c}\right)^2}{\gamma^2}} \right].$$

For

$$\frac{v_x}{c} \ll \frac{\delta}{\omega_0},$$

$$\beta = -\frac{8\hbar\omega_0^2 \,\delta s_0}{\gamma c^2 \left[1+4\left(\dfrac{\delta}{\gamma}\right)^2\right]^2}.$$

For $\beta > 0$, one needs

$$\delta < 0.$$

(c) Characteristic de-Broglie wavelength at temperature T equals $\lambda =$

$\dfrac{\hbar}{\sqrt{mk_B T}}$. To consider the atoms as point-like objects one needs this distance

to be much smaller than characteristic inter particle separation $n^{-\frac{1}{3}}$. From the condition that these two lengths are of the same order of magnitude we get

$$T_Q = \frac{\hbar^2 n^{\frac{2}{3}}}{k_B m} \approx 10^{-6} \, \text{K}.$$

(d) $\langle \Delta p^2 \rangle = \dfrac{\hbar^2 \omega_0^2}{c^2} \approx 10^{-54} \, \text{kg}^2 \, \text{m}^2 / \text{s}^2$ — this is the mean square recoil momentum of a photon.

(e) Assume that the steady state value of the average square of the momentum of atom equals P_0^2. In steady state regime this quantity does not change with time, and temperature is obtained according to $3k_B \dfrac{T_d}{2} = \dfrac{P_0^2}{2m}$. Let the momentum at some point of time in steady state regime be P_0. Let us consider the value of the momentum after some time t. During this time the atom will participate in $N = 6\gamma_p t \gg 1$ absorption-emission processes (6 comes from the number of lasers). For each absorption-emission event the atom gets two recoil kicks, each with a mean square value $\langle \Delta p^2 \rangle$ calculated in part d) (one kick is during absorption and one is during emission). The directions of these kicks are uncorrelated for different events, so this leads to an increase of the mean square of the momentum by $2N \langle \Delta p^2 \rangle$.

On the other hand, atoms are cooled because of the dissipative force, and the change of the mean square of the momentum because of this process is $-\dfrac{2\beta P_0^2 t}{m}$. For steady state solution these two processes compensate each other, so we obtain:

$$P_0^2 = 12 \langle \Delta p^2 \rangle \gamma_p \frac{m}{2\beta} = \frac{3\hbar m \gamma \left(\dfrac{2 \mid \delta \mid}{\gamma} + \dfrac{\gamma}{2 \mid \delta \mid} \right)}{4}.$$

Thus the temperature

$$T_d = \frac{\hbar\gamma\left(\frac{2\,|\,\delta\,|}{\gamma} + \frac{\gamma}{2\,|\,\delta\,|}\right)}{4k_B}.$$

(f) The minimum is achieved for $\delta = -\frac{\gamma}{2}$, and equals $\frac{\hbar\gamma}{2k_B} = 2.4 \cdot 10^{-4}\,\mathrm{K}$.

Experimental Competition

April 26, 2006   Time available: 5 hours

Part 1

Measurement of the specific heat of aluminum in the 45℃ – 65℃ temperature range.

In this part, you can use the following equipment ONLY:

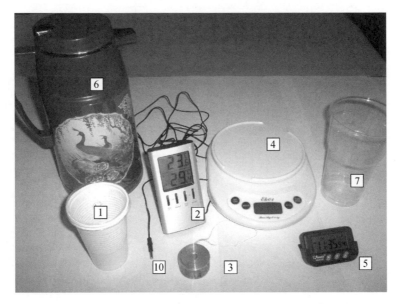

Fig. 7 – 9.

(1) A plastic cup with a cap which has a hole for the thermometer.

(2) A digital thermometer with accuracy of 0.1℃. The temperature of the sensor (10) is shown on the top display, and bottom display shows the temperature of the room. Do not press max/min button; pressing max/min button changes the readings between current, maximal and minimal values. If water temperature exceeds 70℃, the thermometer shows "H" denoting it is out of its range.

Warning: Do Not Use the Thermometer to Measure the Temperature of Liquid Nitrogen! Thermometer Can Be Used Only in Part 1.

(3) An aluminum cylinder with a hole.

(4) Electronic scales with accuracy of 1 g. Make sure that the scales are situated on the flat surface. The button Tare/Zero serves as On/Off and sets the zero reading of the scales. Do not press any other buttons. Note: the scales automatically turn OFF after some time; you have to turn them ON back and reset the zero reading of the scales.

(5) A digital timer. Pressing left button shifts the timer from Clock mode to Stopwatch mode. In Stopwatch mode the middle button serves as Stop/Start, and the right button serves as Reset.

(6) Dewar flask with hot water.

(7) A jar for the used water.

(8) Plotting (graphing) paper (2 pages).

(9) Pieces of thread.

The results of the measurement of specific heat of aluminum will be used in Part 2 of the experimental problem.

The specific heat of aluminum should be determined from the comparison of two experimental curves:

1) the cooling curve of hot water in a plastic cup without the aluminum cylinder (the first experiment);

2) the cooling curve of hot water in a plastic cup with aluminum cylinder immersed (the second experiment).

The specific heat of water is given $c_w = 4.20$ kJ/(kg K).

Density of water $\rho_w = 1.00 \cdot 10^3$ kg/m^3.

Density of aluminum $\rho_{Al} = 2.70 \cdot 10^3$ kg/m^3.

Warning: Be very careful with hot water. Remember that water at temperature $T > 50$℃ can cause burns. Do Not Use Liquiid Nitrogen in This Part!

The task

(1a) Derive theoretically an expression for aluminum specific heat c_{Al} in terms of experimentally measured quantities: mass m_1 of hot water in the first experiment, mass m_2 of hot water in the second experiment, mass m of aluminum cylinder and the ratio of heat capacities $K = \dfrac{C_1}{C_2}$, where C_1 is the heat capacity of water in the first experiment, C_2 is the combined heat capacity of water and aluminum cylinder in the second experiment.

In parts (1b) and (1c) you will perform measurements to determine K. Parts (1b) and (1c) should be performed with closed caps. Assume that in this case heat exchange of the contents of the cup with the environment depends linearly on the difference in their temperatures. The linearity coefficient depends only on the level of the water in the cup. Make sure that the aluminum cylinder is fully immersed into water in part (1c). You can neglect the heat capacity of the cup.

(1b) Perform the first experiment — investigate the relationship between the temperature of water T_1 and time t in the range of temperatures from 45℃ to 65℃. Provide a table of measurements. Write the value of m_1 on the answer sheet.

(1c) Perform the second experiment — investigate the relationship between the temperature of water with aluminum cylinder T_2 and time t in the range of temperatures from 45℃ to 65℃. Provide a table of measurements. Write the values of m_2 and m on the answer sheet.

(1d) Use graphs to determine the ratio of the heat capacities $K = \dfrac{C_1}{C_2}$ and the uncertainty ΔK. Write the values of K and ΔK on the answer sheet.

(1e) Determine the numerical value of c_{Al} and estimate the uncertainty of measurement Δc_{Al}. Write the values of c_{Al} and Δc_{Al} on the answer sheet.

Solutions

(1a) The heat capacity of water in the first experiment is

$$C_1 = c_w m_1,\tag{1}$$

where $c_w = 4.2 \text{ kJ}/(\text{kg K})$ is the specific heat of water and m_1 is the mass of water in the experiment. The heat capacity of water with the aluminum cylinder immersed in it:

$$C_2 = c_w m_2 + c_{Al} m, \tag{2}$$

where m_2 is the mass of water in the experiment, c_{Al} is the specific heat of aluminum and m is the mass of aluminum cylinder. The ratio of heat capacities is $K = \dfrac{C_1}{C_2}$. Then, specific heat capacity of aluminum is determined by the formula

$$c_{Al} = \frac{m_1 - K \cdot m_2}{K \cdot m} c_w. \tag{3}$$

The ratio of heat capacities K can be determined from the experiment in (1b) and (1c). To be able to extract K from these two experiments, one should perform the measurements in the regime such that the level of water is the same for both experiments. This can be done by marking the level of the water on the side of the cup by usual pen, or by choosing the masses of water such that the equation

$$\frac{m_1}{\rho_w} = \frac{m_2}{\rho_w} + \frac{m}{\rho_{Al}}$$

is closely satisfied. Since Eq. (3) have the difference in the numerator, best results are obtained when m_1 is chosen to be close to m.

(1b) The following table shows the T_1, — the temperature of hot water as it cools down, as a function of time t in the 45°C – 65°C temperature range:

N	T_1 (°C)	t(min. sec)	t(min)	$\ln(T_1 - T_r)$
1	65	0. 00	0.0	3.72
2	64	0. 27	0.5	3.69
3	63	0. 56	0.9	3.67
4	62	1. 49	1.8	3.64
5	61	2. 08	2.1	3.62
6	60	3. 02	3.0	3.59
7	59	3. 48	3.8	3.56

Cont.

N	$T_1(°C)$	$t(\text{min. sec})$	$t(\text{min})$	$\ln(T_1 - T_r)$
8	58	4. 34	4.6	3.53
9	57	5. 20	5.3	3.51
10	56	5. 48	5.8	3.47
11	55	6. 33	6.6	3.44
12	54	7. 37	7.6	3.41
13	53	8. 50	8.8	3.38
14	52	9. 37	9.6	3.34
15	51	10. 13	10.2	3.31
16	50	11. 26	11.4	3.27
17	49	12. 39	12.7	3.23
18	48	13. 52	13.9	3.19
19	47	15. 33	15.6	3.15
20	46	16. 17	16.3	3.11
21	45	18. 00	18.0	3.06

The mass of water is $m_1 = (50 \pm 1)$ g and the room temperature is $T_r = (23, 4 \pm 0, 2)°C$.

(1c) The following table shows the temperature T_2 of hot water with aluminum cylinder immersed as the water cools down as a function of time t in the $45°C - 65°C$ temperature range:

N	$T_1(°C)$	$t(\text{min. sec})$	$t(\text{min})$	$\ln(T_1 - T_r)$
1	65	0. 00	0.0	3.72
2	64	0. 18	0.3	3.69
3	63	0. 46	0.8	3.67
4	62	1. 32	1.5	3.64
5	61	2. 00	2.0	3.62
6	60	2. 26	2.4	3.59

Cont.

N	T_1(℃)	t(min. sec)	t(min)	$\ln(T_1 - T_r)$
7	59	3. 03	3.0	3.56
8	58	3. 39	3.7	3.53
9	57	4. 25	4.4	3.51
10	56	4. 45	4.8	3.47
11	55	5. 29	5.5	3.44
12	54	6. 24	6.4	3.41
13	53	7. 19	7.3	3.38
14	52	8. 05	8.1	3.34
15	51	8. 33	8.5	3.31
16	50	9. 27	9.5	3.27
17	49	10. 31	10.5	3.23
18	48	11. 35	11.5	3.19
19	47	12. 58	13.0	3.15
20	46	13. 35	13.6	3.11
21	45	14. 57	15.0	3.06

The mass of aluminum cylinder is $m = (69 \pm 1)$g, the mass of water is $m_2 = (27 \pm 1)$g, and the room temperature is $T_r = (23, 4 \pm 0, 2)$℃.

The graphs $T_1(t)$ and $T_2(t)$ are shown below:

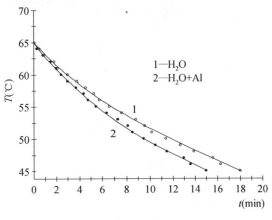

Fig. 7 - 10.

(1d) Water in the first experiment and water with aluminum cylinder immersed in the second experiment cool down because of heat exchange with the air in the room according to the following linear law:

$$\alpha(T - T_r)dt = -CdT, \tag{4}$$

where α is a constant and C is heat capacity (C_1 or C_2). If we integrate the expression (4), we get

$$T - T_r = Ae^{-\frac{\alpha}{C}t}, \tag{5}$$

where $A = T_0 - T_r$ (T_0 is the initial temperature of water in the experiment from the experimental data).

Various methods can be used to obtain the ratio of heat capacities but the most precise result can be obtained from the following linear relation:

$$\ln(T - T_r) = \ln A - \frac{\alpha}{C}t. \tag{6}$$

The graphs $\ln[T_1(t) - T_r]$ and $\ln[T_2(t) - T_r]$ appear to be approximately linear and are shown below:

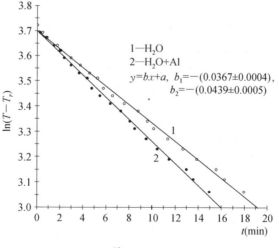

$1—H_2O$
$2—H_2O+Al$
$y = bx + a$, $b_1 = -(0.0367 \pm 0.0004)$,
$b_2 = -(0.0439 \pm 0.0005)$

Fig. 7 - 11.

We can obtain the ratio $K = \dfrac{C_1}{C_2}$ by comparing the slopes of the graphs derived from the first and second experiments. The value of K obtained in terms of the slopes of the two linear relationships is as follows:

$$K = \frac{b_2}{b_1} = \frac{-0.0439}{-0.0367} = 1.196.$$

And the uncertainty is:

$$\Delta K = K \cdot \left(\frac{\Delta b_1}{b_1} + \frac{\Delta b_2}{b_2} \right) = 1.196 \cdot \left(\frac{0.0004}{0.0367} + \frac{0.0005}{0.0439} \right)$$
$$= 1.196 \times 0.022 = 0.026, \quad \varepsilon_K = 2\%.$$

(1e) At $K = 1.196$ the specific heat of aluminum obtained from formula (3) is:

$$c_{Al} = 0.90 \text{ kJ}/(\text{kg} \cdot \text{K}).$$

And the uncertainty is:

$$\Delta c_{Al} = c_{Al} \left(\frac{\Delta m_1 + K \Delta m_2}{m_1 - K m_2} + \frac{\Delta m}{m} + \frac{\Delta K}{K} \frac{m_1}{m_1 - K m_2} \right)$$
$$= 0.90 \cdot \left(\frac{2.196 \text{ g}}{17.7 \text{ g}} + \frac{1 \text{ g}}{69 \text{ g}} + \frac{50 \text{ g}}{17.7 \text{ g}} \frac{0.026}{1.196} \right)$$
$$= 0.90 \cdot 0.2 = 0.18 \text{ kJ}/(\text{kg} \cdot \text{K}).$$

Part 2

Measurement of the Specific Latent Heat of Evaporation of Liquid Nitrogen

In this part you can use the following equipment:

(1) A Styrofoam cup with a cap.

(2) Dewar flask with liquid nitrogen.

(3) An aluminum cylinder with a hole (item 3, Part 1).

(4) Electronic scales with accuracy of 1 g (item 4, Part 1).

(5) A digital timer (item 5, Part 1).

(6) Plotting (graphing) paper (2 pages).

(7) Pieces of thread.

The specific latent heat of evaporation of water is well known, while

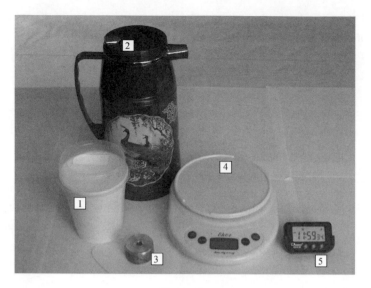

Fig. 7 - 12.

rarely we have to deal with one the main atmospheric gases, nitrogen, in its liquid form. The boiling temperature of liquid nitrogen under normal atmospheric pressure is very low, $T_N = 77 \, \text{K} = -196 \, ^\circ\text{C}$.

In this experiment you are asked to measure the specific latent heat of evaporation of nitrogen. Because of heat exchange with the environment nitrogen in a Styrofoam cup evaporates and its mass decreases at some rate. When an aluminum cylinder initially at room temperature is immersed into nitrogen, nitrogen will boil violently until the temperature of the aluminum sample reaches the temperature of liquid nitrogen. The final brief ejection of some amount of vaporized nitrogen from the cup indicates that aluminum has stopped cooling — this ejection is caused by the disappearance of the vapor layer between aluminum and nitrogen. After aluminum reaches the temperature of nitrogen, the evaporation of nitrogen will continue.

When considering a wide range of temperatures, one can observe that the specific heat of aluminum c_{Al} depends on absolute temperature. The graph of aluminum's specific heat in arbitrary units versus temperature is shown in Fig. 7 - 13. Use the result of specific heat measurement in 45℃ - 65℃ temperature range in Part 1 to normalize this curve in absolute units.

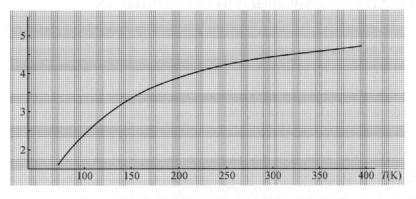

Fig. 7 - 13. The relationship between aluminum's specific heat in arbitrary units and temperature.

Warning: (1) Liquid nitrogen has temperature $T_N = -196℃$. To prevent frostbite do not touch nitrogen or items which were in contact with nitrogen. Make sure to keep away your personal metal belongings such as jewelry, wrist watch, etc. .

(2) Do not put any irrelevant items into nitrogen.

(3) Be careful while putting the aluminum cylinder into the liquid nitrogen to prevent spurts or spilling.

The task

(2a) Measure the evaporation rate of nitrogen in a Styrofoam cup with a closed cap, and measure the mass of nitrogen evaporated during the cooling of the aluminum cylinder (aluminum cylinder is loaded through a hole in the cup). Proceed in the following manner. Set up the Styrofoam on the scales, pour about 250 g of liquid nitrogen in it, wait about 5 minutes and then start taking measurements. After some amount of nitrogen evaporates, immerse the aluminum cylinder into the cup — this will result in a violent boiling. After aluminum cylinder cools down to the temperature of nitrogen, evaporation calms down. You should continue taking measurements in this regime for about 5 minutes until some additional amount of nitrogen evaporates. During the whole process record the readings of the scales $M(t)$

as a function of time.

IN NO CASE SHOULD YOU TOUCH THE ALUMINUM CYLINDER AFTER IT WAS SUBMERGED INTO LIQUID NITROGEN.

In your report provide a table of $M(t)$ and $m_N(t)$, where $m_N(t)$ is the mass of the evaporated nitrogen.

(2b) Using the results of the measurements of $M(t)$ in (2a), plot the graph of the mass of evaporated nitrogen m_N versus time t. The graph should illustrate all the three stages of the process — the calm periods before and after immersion of aluminum, and the violent boiling of nitrogen.

(2c) Determine from the graph the mass m_N^{Al} of nitrogen evaporated only due to heat exchange with the aluminum cylinder, as it is cooled down from room temperature to the temperature of liquid nitrogen. In order to do this you have to take into account the heat exchange with the environment through the cup before, during and after the cooling of aluminum. Write the value of m_N^{Al} and its uncertainty Δm_N^{Al} on the answer sheet.

(2d) Using the result of measurement of aluminum's specific heat in the temperature range of $45\,°C - 65\,°C$ (Part 1), normalize the graph of the relationship between aluminum's specific heat and temperature from arbitrary to absolute units. On the answer sheet write the value of the coefficient β of conversion from arbitrary units to absolute units:

$$c_{Al}(J/(kg \cdot K)) = \beta \cdot c_{Al}(\text{arb. units}).$$

(2e) Using the results of measurement of the mass of nitrogen evaporated due to cooling of the aluminum cylinder and the normalized graph of the relationship between specific heat and temperature, determine nitrogen's specific latent heat of evaporation λ. Write the value of λ and its uncertainty $\Delta\lambda$ on the answer sheet.

🔑 Solutions

(2a) Below is the table showing the the mass of evaporated nitrogen m_N versus time:

N	$M(g)$	t(min. sec)	t(min)	$m_N(g)$
1	250	0. 0	0	0
2	248	0. 19	0.3	2
3	246	0. 39	0.6	4
4	244	0. 59	1	6
5	242	1. 19	1.3	8
6	240	1. 38	1.6	10
7	238	2. 00	2	12
8	236	2. 19	2.3	14
9	234	2. 41	2.7	16
10	232	3. 02	3	18
11	230	3. 22	3.4	20
12	228	3. 44	3.7	22
13	226	4. 06	4.1	24
14	224	4. 28	4.5	26
15	222	4. 50	4.8	28
16	220	5. 13	5.2	30
17	274	5. 52	5.9	45
18	269	6. 00	6	50
19	264	6. 07	6.1	55
20	259	6. 18	6.3	60
21	254	6. 30	6.5	65
22	249	6. 41	6.7	70
23	244	6. 54	6.9	75
24	239	7. 09	7.1	80
25	234	7. 25	7.4	85
26	229	7. 40	7.7	90
27	224	7. 48	7.8	95
28	222	8. 06	8.1	97

Cont.

N	M(g)	t(min. sec)	t(min)	m_N(g)
29	219	8. 49	8.8	100
30	217	9. 16	9.3	102
31	215	9. 44	9.7	104
32	213	10. 14	10.2	106
33	211	10. 44	10.7	108
34	209	11. 14	11.2	110
35	207	11. 45	11.7	112
36	205	12. 13	12.2	114
37	203	12. 43	12.7	116
38	201	13. 14	13.2	118
39	199	13. 46	13.7	120
40	197	14. 16	14.3	122
41	195	14. 50	14.8	124

(2b) Below is the graph of the mass of evaporated nitrogen $m_N(t)$ versus time t (all the three stages of the experiment are shown):

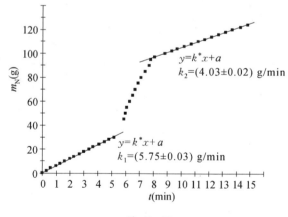

$$y=k^*x+a$$
$$k_2=(4.03\pm0.02)\ \text{g/min}$$

$$y=k^*x+a$$
$$k_1=(5.75\pm0.03)\ \text{g/min}$$

Fig. 7 – 14.

(2c) Applying the ordinary least squares method to Fig. 7 – 14, we can determine nitrogen's evaporation rates k_1 and k_2, before the immersion of aluminum and after violent boiling, respectively. Nitrogen's evaporation rate before the immersion of aluminum is

$$k_1 = (5.75 \pm 0.03) \text{g/min.}$$

And the evaporation rate after violent boiling ends is

$$k_2 = (4.03 \pm 0.02) \text{g/min.}$$

It is obvious from these rates that evaporation rate depends on the amount of nitrogen in the cup. Therefore the evaporation rate during violent boiling due to heat exchange with the environment can be estimated as the average of the evaporation rate before and after violent boiling:

$$k = \frac{k_1 + k_2}{2} = \frac{5.75 + 4.03}{2} = 4.89 \text{ g/min.}$$

To determine the mass m_N^{Al} consider the time period from $t_1 = 5.2$ min to $t_2 = 7.8$ min. t_1 is set by the moment of immersion, t_2 is determined as described in the introduction, or by analyzing the $m_N(t)$ dependence. Then

$$\begin{aligned}
m_N^{Al} &= (m_N(t_2) - m_N(t_1)) - k(t_2 - t_1) \\
&= (95 - 30) - 4.89 \times (7.8 - 5.2) \\
&= 52.3 \text{ g.}
\end{aligned}$$

It should be noted that any other mean, for instance geometric mean, can be used as an estimate of nitrogen's average evaporation rate due to heat exchange with the environment. The difference between the arithmetic mean and geometric mean will be used as an estimate of the error of average evaporation rate: $\Delta k = \pm 0.10$ g/min.

The uncertainty is

$$\Delta m_N^{Al} = \Delta m_N(t_2) + \Delta m_N(t_1) + k(t_2 - t_1)\left(\frac{\Delta k}{k} + 2\frac{\Delta t}{t}\right) = 2.4 \text{ g.}$$

Then

$$m_N^{Al} = (52.3 \pm 2.4) \text{g.}$$

(2d) In the 45℃ – 65℃ temperature range the specific heat of aluminum is approximately constant and equal to the measurement performed in

Part 1:

$$c_{Al} = 0.90 \text{ kJ/(kg} \cdot \text{K)}.$$

In this temperature range, the value of specific heat in arbitrary units is

$$c_{Al}(\text{arb. units}) = 4.5 \text{ arb. units}.$$

Consequently the coefficient of conversion of specific heat from arbitrary units to absolute units, β, is

$$\beta = \frac{c_{Al}}{c_{Al}(\text{arb. units})} = \frac{0.90}{4.5} = 0.2 \text{ kJ/(kg} \cdot \text{K} \cdot \text{arb. units)}.$$

(2e) The amount of heat transferred from the aluminum cylinder to the liquid nitrogen as the cylinder is cooled down to the temperature $T_N = 77 \text{ K}$, is equal to

$$Q = m \cdot \int_{T_N}^{T_r} c_{Al}(T) dT.$$

The value of this integral can be found using numerical integration. It can be approximated as the area under the $c_{Al}(T)$ curve. In our experiment, the number of cells under the curve is $N = 311 \pm 1$ and each cell represents $0.5 \text{ J/(g} \cdot \text{K)}$, then

$$\int_{T_N}^{T_r} c_{Al}(T) dt = 155.5 \text{ J/g}.$$

Then the amount of heat released is

$$Q = 69 \times 155.5 = 10.73 \text{ kJ}.$$

The value of specific latent heat of nitrogen's evaporation can be found from the heat balance equation,

$$m_N^{Al} \cdot \lambda = Q.$$

Thus we finally get

$$\lambda = \frac{Q}{m_N^{Al}} = \frac{10.3 \text{ kJ}}{52.3 \text{ g}} = 205 \text{ J/g}$$

$$\Delta\lambda = \lambda \left(\frac{\Delta c_{Al}}{c_{Al}} + \frac{\Delta N}{N} + \frac{\Delta m_N^{Al}}{m_N^{Al}} + \frac{\Delta m}{m} \right) = 60 \text{ J/g}.$$

Minutes of the Eighth Asian Physics Olympiad

Shanghai (China), April 21 - 29, 2007

1. The following 22 countries and regions were present at the 8th Asian Physics Olympiad:

Australia (8 + 2), Azerbaijan (2 + 1), Brunei Darussalam (6 + 2), Cambodia (6 + 2), China (team I : 8 + 2, team II : 8 + 1), Chinese Taipei (8 + 2), Hong Kong (8+2), India (2+1), Indonesia (team I : 8+2, team II : 8+2), Israel (8+ 2), Kazakhstan (5+2), Kyrgyzstan (7+2), Laos (4+1), Macau (4+1), Mongolia (8 + 2), Nepal (5 + 2), Sri Lanka (7 + 2), Singapore (8 + 2), Tajikistan (2 + 2), Thailand (8 + 2), Turkmenistan (8 + 1), Vietnam (7 + 2). The first number in parentheses denotes the number of the competitors while the second number denotes the number of the leaders.

In addition, Japan was represented by three observers.

2. Results of marking the papers by the Organizers were presented:

The best score was 43.3 points achieved by Yun Yang from China (Absolute winner of the 8th APhO). The second best score (43.2 points) was achieved by Muhammad Firmansyah from Indonesia. The third best score (43.1 points) was achieved by Han Yu Zhu from China. The following criteria for awarding the medals and honorable mentions were established according to the Statutes:

Gold Medal:	38 points,
Silver Medal:	33 points,
Bronze Medal:	28 points,
Honorable Mention:	21 points.

According to the above criteria 13 Gold Medals, 10 Silver Medals, 13 Bronze Medals and 25 Honorable Mentions were awarded.

The results of the 8th APhO have been distributed to all the delegations.

3. In addition to the regular prizes, the following special prizes were awarded:

● for the Absolute Winner: Yun Yang (China)

● for the best score in the theoretical part of the competition: Han Yu Zhu (China)

● for the best score in the experimental part of the competition: Chang-Chi Wu (Chinese Taipei)

- for the best score among the female participants: Kathryn Zealand (Australia)
- for the best new comer (President's Prize): Buddhi Wijerathna (Sri Lanka)
- for the First Time of Entering APhO (for Teams):

  Brunei

  Hong Kong

  Macau

  Nepal

  Sri Lanka

 4. President and Secretary of the APhOs presented the provisional list of hosting the future competitions:

- 2008 – Mongolia (invitations made)
- 2009 – Thailand (confirmed)
- 2010 – Chinese Taipei (confirmed)
- 2011 – Israel (confirmed)
- 2012 – Cambodia (preliminary contact)

 5. The International Board discussed the proposal made by the Honorary President of the APhOs, Dr. Gorzkowski, suggesting the change of the item #5 of the APhO Statutes which referred to inviting President, Secretary, Honorary President, and Honorary Members to each Asian Physics Olympiad at cost (including travel expenses) of the organizing country. The proposal was not accepted: 23 votes in favor, while the necessary minimum (qualified $\frac{2}{3}$-majority of votes) was 28.

 6. President of the APhOs, acting on behalf of all the participants, expressed deep thanks to China Association for Science and Technology, Shanghai Municipal Government, Professor Zuimin Jiang and his collaborators for excellent preparing and executing of the 8th Asian Physics Olympiad. Deep thanks were also conveyed to Ministry of Education of China, National Natural Science Foundation of China, Shanghai Association for Science and Technology, Fudan University, Chinese Physical Society, and Shanghai Physics Society.

 7. Acting on behalf of the Organizers of the next Asian Physics Olympiad, President of National University of Mongolia, Professor Tserensodnom Gantsog,

announced that the 9th APhO would be held in Ulaanbaatar, Mongolia on April 20 – 28, 2008 and cordially invited all of the participating countries to attend the competition.

Shanghai, China, April 28, 2007

Prof. Yohanes Surya President
of the APhOs

Prof. Ming-Juey Lin
Secretary of the APhOs

Prof. Zuimin Jiang
Executive Chairman Organizing
Committee of the 8th APhO

Dr. Waldemar Gorzkowski
Honorary President of the APhOs

Theoretical Competition

April 23,2007   Time available: 5 hours

Problem 1
Back-and-Forth Rolling of a Liquid-Filled Sphere

Consider a sphere filled with liquid inside rolling back and forth at the bottom of a spherical bowl. That is, the sphere is periodically changing its translational and rotational direction. Due to the viscosity of the liquid inside, the movement of the sphere would be very complicated and hard to deal with. However, a simplified model presented here would be beneficial to the solution of such a problem.

Assume that a rigid thin spherical shell of radius r and mass m is fully filled with some liquid substance of mass M, denoted as W. W has such a unique property that usually it behaves like an ideal liquid, i. e. without any viscosity, while in response to some special external influence (such as electric field), it transits to solid state immediately with the same volume; and once the applied influence removed, the liquid state recovers immediately. Besides, this influence does not give rise to any force or torque exerting on the sphere. This liquid-filled spherical shell (for convenience, called "the sphere" hereafter) is supposed to roll back and forth at the bottom of a spherical bowl of radius R $(R > r)$ without any relative slipping, as shown in the figure. Assume the sphere moves only in the vertical plane (namely, the plane of the figure), please study the movement of the sphere for the following three cases:

1. W behaves as in ideal solid state, meanwhile W contacts the inner wall of the spherical shell so closely that they can be taken as solid sphere as a whole of radius r with an abrupt density change across the interface between the inside wall of the shell and W.

(1) Calculate the rotational inertia I of the sphere with respect to the axis passing through its center C. (You are asked to show detailed steps.)

(2) Calculate the period T_1 of the sphere rolling back and forth with a small amplitude without slipping at the bottom of the spherical bowl.

2. W behaves as ideal liquid with no friction between W and the spherical shell. Calculate the period T_2 of the sphere rolling back and forth with a small amplitude without slipping at the bottom of the spherical bowl.

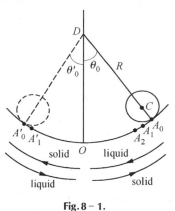

Fig. 8 - 1.

3. W transits between ideal solid state and ideal liquid state.

Assume at time $t = 0$, the sphere is kept at rest, the line CD makes an angle θ_0 ($\theta_0 \ll 1$ rad) with the plumb line OD, where D is the center of the spherical bowl. The sphere contacts the inner wall of the bowl at point A_0, as shown in the figure. Release the sphere, it starts to roll left from rest. During the motion of the sphere from A_0 to its equilibrium position O, W behaves as ideal liquid. At the moment that the sphere passes through point O, W changes suddenly into solid state and sticks itself firmly on the inside wall of the sphere shell until the sphere reaches the left highest position A'_0. Once the sphere reaches A'_0, W changes suddenly back into the liquid state. Then, the sphere rolls right; and W changes suddenly into solid state and sticks itself firmly on the inside wall of the spherical shell again when the sphere passes through the equilibrium position O. When the sphere reaches the right highest position A_1, W changes into liquid state once again. Then the whole circle repeats time after time. The sphere rolls right and left periodically but with the angular amplitude decreased time after time. The motion direction of the sphere is shown by curved arrows in the figure, together with the words "solid" and "liquid" showing corresponding state of W. It is assumed that during such process of rolling back and forth, no any relative slide happens between the sphere and the inside wall of the bowl (or, alternatively, the bottom of the bowl can supply as enough friction as needed). Calculate the period T_3 of the sphere rolling right and left, and the angular amplitude θ_n of the center of the sphere, namely, the angle that the line CD makes with the vertical line OD when the sphere reaches the right highest position A_n for the n-th time (only A_2 is shown in the figure).

🔑 Solution

1. (1) Let I_1 and I_2 denote the rotational inertia of the spherical shell and W in solid state respectively, while I be the sum of I_1 and I_2. The surface mass density of the spherical shell is $\sigma = \dfrac{m}{4\pi r^2}$. Cut a narrow zone from the spherical shell perpendicular to its diameter, which spans a small angle $d\alpha$ with respect to the center of the sphere C, while the spherical zone makes an angle α with the diameter of the spherical shell, which is called C axes hereafter, as shown in Fig. 8 – 2. The rotational inertia of the narrow zone about the C axis is $2\pi r\sin\ \alpha\ (r d\alpha)\ \sigma\ (r\sin\ \alpha)^2$, therefore integral over the whole spherical shell gives

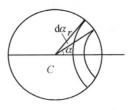

Fig. 8 – 2.

$$I_1 = \int_0^\pi 2\pi r\sin\ \alpha(r d\alpha)\sigma(r\sin\ \alpha)^2 = \frac{2}{3}mr^2. \qquad (1.1)$$

The volume density of W is $\rho = \dfrac{M}{\dfrac{4\pi r^3}{3}}$. By using above result for the spherical zone it can be seen that the rotational inertia of the solid W about the C axis is

$$I_2 = \int_0^r \frac{2}{3}r'^2 \cdot \rho \cdot 4\pi r'^2\, dr' = \frac{2}{5}Mr^2. \qquad (1.2)$$

Then, $$I = I_1 + I_2 = \frac{2}{3}mr^2 + \frac{2}{5}Mr^2. \qquad (1.3)$$

(2) According to the Newton's second law, we can derive the translational motion equation of the center of mass for the sphere along the tangent of the bowl,

$$(m + M)(R - r)\ddot{\theta} = -(m + M)g\theta + f, \qquad (1.4)$$

where θ ($\theta \ll 1$) denotes the angular position of the center of mass of the sphere as shown in Fig. 8‑3, and f is the frictional force acting on the sphere by the inside wall of the bowl. From the rotational dynamics, we have

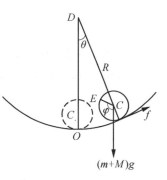

$$fr = -I\ddot{\varphi} = -\left(\frac{2}{3}mr^2 + \frac{2}{5}Mr^2\right)\ddot{\varphi},$$

$$(1.5)$$

Fig. 8‑3.

where φ is the angular position of the reference radius CE with respect to the starting position. Assumed constraint of pure rolling on the motion of the sphere reads,

$$(R - r)\ddot{\theta} = r\ddot{\varphi}, \tag{1.6}$$

Eqs. (1.4)–(1.6) lead to

$$\left(\frac{5}{3}m + \frac{7}{5}M\right)(R - r)\ddot{\theta} = -(m + M)g\theta.$$

This is a motion equation of the type of simple harmonic oscillator. Therefore, we obtain the angular frequency and period of the sphere rolling right and left:

$$\omega_1 = \sqrt{\frac{m + M}{\frac{5m}{3} + \frac{7M}{5}} \frac{g}{R - r}}, \tag{1.7}$$

$$T_1 = 2\pi\sqrt{\frac{R - r}{g}}\sqrt{\frac{\frac{5m}{3} + \frac{7M}{5}}{m + M}}. \tag{1.8}$$

2. This case can be treated similarly, except taking that the ideal liquid does not rotate into consideration. Therefore Eqs. (1.4) and (1.6) are still applicable, while Eq. (1.5) needs to be modified as

$$fr = -I_1\ddot{\varphi} = -\frac{2}{3}mr^2\ddot{\varphi}. \tag{1.9}$$

Eqs. (1.4), (1.6) and (1.9) result in

$$\left(\frac{5m}{3} + M\right)(R - r)\ddot{\theta} = -(m + M)g\theta.$$

Then, the angular frequency and period of the sphere rolling back-and-forth are obtained respectively.

$$\omega_2 = \sqrt{\frac{m + M}{\frac{5m}{3} + M}} \sqrt{\frac{g}{R - r}}, \tag{1.10}$$

$$T_2 = 2\pi \sqrt{\frac{R - r}{g}} \sqrt{\frac{\frac{5m}{3} + M}{m + M}}. \tag{1.11}$$

3. The time taken by the sphere from position A_0 to equilibrium position O is $\frac{T_2}{4}$, $\frac{T_1}{4}$ from O to A_0', and $\frac{T_2}{4}$ from A_0' to O, $\frac{T_1}{4}$ from O to A_1. Although the angular amplitude decreases step by step (see below) during the rolling process of the sphere right and left, the period keeps unchanged. This means

$$T_3 = \frac{1}{2}(T_1 + T_2) = \pi\sqrt{\frac{R - r}{g}} \left(\sqrt{\frac{\frac{5m}{3} + \frac{7M}{5}}{m + M}} + \sqrt{\frac{\frac{5m}{3} + M}{m + M}} \right). \tag{1.12}$$

Next, we calculate the change of the angular amplitude. When the sphere passes through the equilibrium position O after it rolled down from the initial position A_0, the velocity of its center is

$$v_C = \omega_2 (R - r)\theta_0 = \sqrt{\frac{m + M}{\frac{5m}{3} + M}} \sqrt{g(R - r)} \, \theta_0. \tag{1.13}$$

Now the angular velocity of the spherical shell rotating about the C axis is

$$\Omega = \frac{v_C}{r} = \frac{\theta_0}{r} \sqrt{\frac{m + M}{\frac{5m}{3} + M}} \sqrt{g(R - r)}, \tag{1.14}$$

where C axis is the axis of rotation through the center of the sphere and perpendicular to the paper plane of Fig. 8 - 2. When W behaves as liquid (before it changes into solid state), the angular momentum of the sphere relative to point O is

$$L = (m + M)v_C r + I_1 \Omega. \tag{1.15}$$

When W changes suddenly into solid state, due to the fact that both gravitational and frictional force pass through point O, the angular momentum of the sphere relative to O is conserved, we have

$$\begin{aligned} L &= (m + M)v_C r + I_1 \Omega \\ &= [I + (m + M)r^2]\Omega', \end{aligned} \tag{1.16}$$

where Ω and Ω' represent the angular velocity of the sphere immediately before and after passing through point O. Therefore

$$\Omega' = \frac{(m + M)v_C r + I_1 \Omega}{I + (m + M)r^2} = \frac{v_C}{r} \frac{\dfrac{5m}{3} + M}{\dfrac{5m}{3} + \dfrac{7M}{5}}, \tag{1.17}$$

while after passing through point O the velocity of the center of the sphere becomes

$$v_C' = \Omega' r = v_C \frac{\dfrac{5m}{3} + M}{\dfrac{5m}{3} + \dfrac{7M}{5}}. \tag{1.18}$$

Once the sphere reaches the left highest position A_0' corresponding to the left angular amplitude θ_0' we have

$$v_C' = \omega_1 (R - r)\theta_0'.$$

However, $$v_C = \omega_2 (R - r)\theta_0.$$

From above two expressions we obtain

$$\theta_0' = \frac{v_C' \omega_2}{v_C \omega_1}\theta_0 = \theta_0 \sqrt{\frac{\dfrac{5m}{3} + M}{\dfrac{5m}{3} + \dfrac{7M}{5}}}. \tag{1.19}$$

Similarly we can treat the process that the sphere rolls from position A_0' back to A_1, the second highest position on the right, corresponding to the second right angular amplitude θ_1 , and obtain

$$\frac{\theta_1}{\theta_0^r} = \frac{\theta_0'}{\theta_0}.$$

Then,
$$\theta_1 = \frac{\theta_0'^2}{\theta_0} = \frac{\dfrac{5m}{3} + M}{\dfrac{5m}{3} + \dfrac{7M}{5}}\theta_0.$$

Following the similar procedure repeatedly we finally obtain:

$$\theta_n = \left(\frac{\dfrac{5m}{3} + M}{\dfrac{5m}{3} + \dfrac{7M}{5}}\right)^n \theta_0. \tag{1.20}$$

Problem 2

2A. Optical Properties of an Unusual Material

The optical properties of a medium are governed by its relative permittivity (ε_r) and relative permeability (μ_r). For conventional materials like water or glass, which are usually optically transparent, both of their ε_r and μ_r are positive, and refraction phenomenon meeting Snell's law occurs when light from air strikes obliquely on the surface of such kind of substances. In 1964, a Russia scientist V. Veselago rigorously proved that a material with simultaneously negative ε_r and μ_r would exhibit many amazing and even unbelievable optical properties. In early 21st century, such unusual optical materials were successfully demonstrated in some laboratories. Nowadays study on such unusual optical materials has become a frontier scientific research field. Through solving several problems in what follows, you can gain some basic understanding of the fundamental optical properties of such unusual materials. It should be noticed that a material with simultaneously negative ε_r and μ_r possesses the following important property. When a light wave propagates forward inside such a medium for a distance Δ, the phase of the light wave will decrease, rather than increase an amount of $\sqrt{\varepsilon_r\mu_r}k\Delta$ as what happens in a conventional medium with simultaneously positive ε_r and μ_r. Here, positive root is always taken when we apply the square-root calculation, while k is the wave vector of the light. In the questions listed below, we assume that both the relative permittivity and permeability of air are equal to 1.

1. (1) According to the property described above, assuming that a light beam strikes from air on the surface of such an unusual material with relative permittivity $\varepsilon_r < 0$ and relative permeability $\mu_r < 0$, prove that the direction of the refracted light beam depicted in Fig. 8 – 4 is reasonable.

(2) Fig. 8 – 4 show the relationship between refraction angle θ_r (the angle that refracted beam makes with the normal of the interface between air and the material) and incidence angle θ_i.

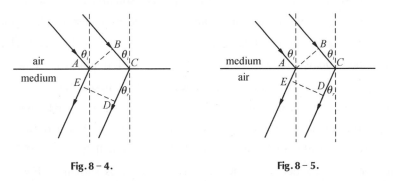

Fig. 8 – 4.   Fig. 8 – 5.

(3) Assuming that a light beam strikes from the unusual material on the interface between it and air, prove that the direction of the refracted light beam depicted in Fig. 8 – 5 is reasonable.

(4) Fig. 8 – 5 show the relationship between the refraction angle θ_r (the angle that refracted beam makes with the normal of the interface between two media) and the incidence angle θ_i.

2. As shown in Fig. 8 – 6 a slab of thickness d, which is made of an unusual optical material with $\varepsilon_r = \mu_r = -1$, is placed in air, with a point light source located in front of the slab separated by a distance of $\frac{3}{4}d$. Accurately draw the ray diagrams for the three light rays radiated from the point source. (Hints: under the conditions given in this problem, no

reflection would happen at the interface between air and the unusual material.)

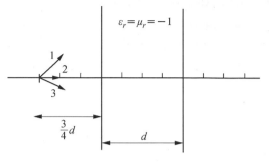

Fig. 8 – 6.

3. As shown in Fig. 8 – 7, a parallel-plate resonance cavity is formed by two plates parallel with each other and separated by a distant d. Optically one of the plates, denoted as Plate 1 in Fig. 8 – 7, is ideally reflective (reflectance equals to 100%), and the other one, denoted by Plate 2, is partially reflective (but with a high reflectance). Suppose plane light waves are radiated from a source located near Plate 1, then such light waves are multiply reflected by the two plates inside the cavity. Since optically the Plate 2 is not ideally reflective, some light waves will leak out of Plate 2 each time the light beam reaches it (ray 1, 2, 3, as shown in Fig. 8 – 7), while some light waves will be reflected by it. If these light waves are in-phase, they will interfere with each other constructively, leading to resonance. We assume that the light wave gains a phase of π by reflection at either of the two plates. Now we insert a slab of thickness $0.4d$ (shown as the shaded area in Fig. 8 – 7), made of an unusual optical material with $\varepsilon_r = \mu_r = -0.5$, into the cavity parallel to the two plates. The remaining space is filled with air inside the cavity. Let us consider only the situation that the light wave travels along the direction perpendicular to the plates (the ray diagram depicted in Fig. 8 – 7 is only a schematic one), calculate all the wavelengths that satisfy the resonance condition of such a cavity. (Hints: under the condition given here, no reflection would occur at the interfaces between air and the unusual material.)

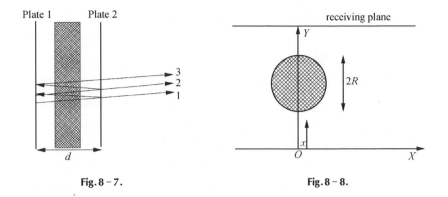

Fig. 8 – 7.   Fig. 8 – 8.

4. An infinitely long cylinder of radius R, made of an unusual optical material with $\varepsilon_r = \mu_r = -1$, is placed in air, its cross section in XOY plane is shown in Fig. 8 – 8 with the center located on Y axis. Suppose a laser source located on the X axis (the position of the source is described by its coordinate x) emits narrow laser light along the Y direction. Show the range of x, for which the light signal emitted from the light source cannot reach the infinite receiving plane on the other side of the cylinder.

2B. Dielectric Spheres Inside an External Electric Field

By immersing a number of small dielectric particles inside a fluid of low-viscosity, you can get the resulting system as a suspension. When an external electric field is applied on the system, the suspending dielectric particles will be polarized with electric dipole moments induced. Within a very short period of time, these polarized particles aggregate together through dipolar interactions so that the effective viscosity of the whole system enhances significantly (the resulting system can be approximately viewed as a solid). This type of phase transition is called "electrorheological effect", and such a system is called "electrorheological fluid" correspondingly. Such an effect can be applied to fabricate braking devices in practice, since the response time of such a phase transition is shorter than conventional mechanism by several orders of magnitude. Through solving several problems in the following, you are given a simplified picture to understand the inherent mechanism of the electrorheological transition.

1. When there are many identical dielectric spheres of radius a

immersed inside the fluid, we assume that the dipole moment of each sphere p, is induced *solely* by the external field E_0, independent of any of the other spheres. (**Note**: $p /\!/ E_0$.)

(1) When two identical small dielectric spheres exist inside the fluid and contact with each other, while the line connecting their centers makes an angle θ with the external field direction (see Fig. 8 − 9), write the expression of the energy of dipole-dipole interaction between the two small contacting dielectric spheres, in terms of p, a and θ. (**Note**: In your calculations each polarized dielectric sphere can be viewed as an electric dipole located at the center of the sphere.)

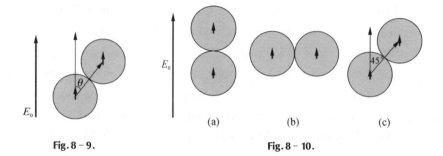

Fig. 8 − 9.   Fig. 8 − 10.

(2) Calculate the dipole-dipole interaction energies for the three configurations shown in Fig. 8 − 10.

(3) Identify which configuration of the system is the most stable one.

(**Note**: In your calculations each polarized dielectric sphere can be viewed as an electric dipole located at the center of the sphere, and the energy of dipole-dipole interaction can be expressed in terms of p and a.)

2. In the case that three identical spheres exist inside the fluid, based on the same assumption as in question 1,

(1) calculate the dipole-dipole interaction energies for the three configurations shown in Fig. 8 − 11;

(2) identify which configuration of the system is the most stable one;

(3) identify which configuration of the system is the most unstable one.

(**Note**: In your calculations each polarized dielectric sphere can be

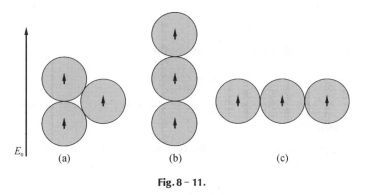

Fig. 8 – 11.

viewed as an electric dipole located at the center of the sphere, and the energy of dipole-dipole interaction can be expressed in terms of p and a.)

 Solution

2A. Optical Properties of an Unusual Material

1. (1)

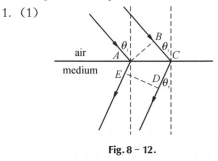

Fig. 8 – 12.

Proof. Assume $E - D$ shows one of the wavefronts of the refracted light. According to the Huygens' principle the phase accumulation from A to E should be equal to that from B via C to D:

$$\phi_{AE} = \phi_{BC} + \phi_{CD}. \tag{2A.1}$$

With the Hints given these phase differences can be calculated respectively, then

$$- \sqrt{\varepsilon_r \mu_r} \, kd_{AE} = kd_{BC} - \sqrt{\varepsilon_r \mu_r} \, kd_{CD}. \tag{2A.2}$$

Simplification of (2A.2) gives

$$-\sqrt{\varepsilon_r\mu_r}\,(d_{AE} - d_{CD}) = d_{BC}. \tag{2A.3}$$

Because $d_{BC} > 0$ and $\sqrt{\varepsilon_r\mu_r} > 0$, we obtain

$$d_{AE} < d_{CD}. \tag{2A.4}$$

Therefore the schematic ray diagram of the refracted light shown in above figure is reasonable.

(2) From the above figure, the refraction angle θ_r and incidence angle θ_i satisfy

$$d_{BC} = d_{AC}\sin\theta_i, \quad d_{CD} - d_{AE} = d_{AC}\sin\theta_r \tag{2A.5}$$

respectively. Substitution of (2A.5) into (2A.3) results in:

$$\sqrt{\varepsilon_r\mu_r}\,\sin\theta_r = \sin\theta_i. \tag{2A.6}$$

(3)

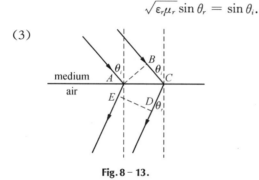

Fig. 8 - 13.

Proof. Assume $E - D$ shows one of the wavefronts of refracted light. According to the Huygens' principle the phase accumulation from A to E should be equal to that from B via C to D:

$$\phi_{AE} = \phi_{BC} + \phi_{CD}. \tag{2A.7}$$

With the Hints given these phase differences can be calculated respectively, then

$$kd_{AE} = -\sqrt{\varepsilon_r\mu_r}\,kd_{BC} + kd_{CD}. \tag{2A.8}$$

Simplification of (2A.8) gives

$$d_{AE} - d_{CD} = -\sqrt{\varepsilon_r\mu_r}\,d_{BC}. \tag{2A.9}$$

Because $d_{BC} > 0$ and $\sqrt{\varepsilon_r\mu_r} > 0$, we obtain

$$d_{AE} < d_{CD}. \tag{2A.10}$$

Therefore the schematic ray diagram of the refracted light shown in the above figure is reasonable.

(4) From the above figure, the refraction angle θ_r and incidence angle θ_i satisfy

$$d_{BC} = d_{AC} \sin \theta_i, \; d_{CD} - d_{AE} = d_{AC} \sin \theta_r \tag{2A.11}$$

respectively. Substitution of (2A.11) into (2A.9) results in:

$$\sin \theta_r = \sqrt{\varepsilon_r \mu_r} \sin \theta_i. \tag{2A.12}$$

2. The ray diagram is shown below.

Illustration:

The light is negatively refracted at both interfaces, and the refraction angle equals to incidence angle. Meanwhile according to the Hints provided there is no reflected light from each interface. Therefore within the medium light rays converge strictly at a point symmetric to the source about the left side of the medium, and on the other side of the medium the rays converge strictly at a point which is symmetric to the image of the source within the medium about the right side of the medium.

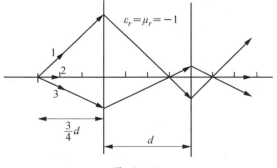

Fig. 8 – 14.

3. The phase difference between the two waves transmitting through the right side of the medium in succession is

$$\Delta\phi = 2k(d - 0.4d) - 2 \cdot 0.5k \cdot 0.4d + 2\pi. \tag{2A.13}$$

On the right side of above equation, the first term shows the phase difference of the light wave accumulated during its propagation in air, the second term shows the phase difference of the light wave accumulated during its propagation in the unusual medium, while the third term accounts for the phase difference of the light wave accumulated due to the two reflections in succession from the interface between air and the medium. Taking $k = \dfrac{2\pi}{\lambda}$, (2A.13) changes into

$$\Delta\phi = 0.8 \frac{2\pi}{\lambda}d + 2\pi. \qquad (2A.14)$$

Resonant condition means

$$0.8 \frac{2\pi}{\lambda}d + 2\pi = m \cdot 2\pi. \qquad (2A.15)$$

Thus
$$\lambda = \frac{0.8d}{m-1}, \quad m = 2, 3, 4, \ldots \qquad (2A.16)$$

4.

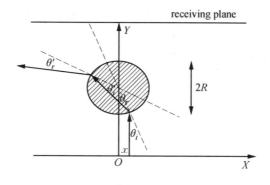

Fig. 8 - 15.

From the given conditions the ray diagram can be accordingly constructed. Above figure shows schematically the ray diagram for $x > 0$. Because for the unusual medium $\varepsilon_r = \mu_r = -1$, from the solution of the first question we have $\theta_i = \theta_r = \theta_i' = \theta_r'$. Therefore the direction of the final outgoing light deviates from that of the incident light by $4\theta_i$. Because the direction of the incident light is given in the y direction, only if the condition

$$\frac{\pi}{2} \leqslant 4\theta_i \leqslant \frac{3\pi}{2} \Rightarrow \frac{\pi}{8} \leqslant \theta_i \leqslant \frac{3\pi}{8} \tag{2A.17}$$

is satisfied the light signal can not reach the receiving plane. Notice

$$\sin \theta_i = \frac{x}{R}, \tag{2A.18}$$

and the similarity of the monotonicity of $\sin \theta$ to that of θ in the range of $\left[0, \frac{\pi}{2}\right]$, we find that (2A.17) goes to

$$\sin\left(\frac{\pi}{8}\right) \leqslant \frac{x}{R} \leqslant \sin\left(\frac{3\pi}{8}\right). \tag{2A.19}$$

Further taking the symmetry about the y axis into consideration we obtain that if the following condition

$$R\sin\left(\frac{\pi}{8}\right) \leqslant |x| \leqslant R\sin\left(\frac{3\pi}{8}\right) \tag{2A.20}$$

is satisfied, the light emitted from a light source located on the x axis can not reach the receiving plane.

2B. Dielectric Spheres Inside an External Electric Field

1. (1) Adopting the polar coordinates, the z component of the electric field produced by a dipole located at the origin with its axis parallel to the z axis is

$$E_z(\theta, \phi) = -\frac{p}{4\pi\varepsilon_0} \frac{1 - 3\cos^2\theta}{r^3} \tag{2B.1}$$

where r is the length of the relative position vector of the two dipoles. In the external electric field E, the energy of a dipole with its axis parallel to the z axis is

$$U = -\boldsymbol{p} \cdot \boldsymbol{E} = -pE_z.$$

Therefore, we obtain that the interaction energy between two contacting small dielectric spheres is

$$U_{12} = \frac{p^2}{4\pi\varepsilon_0} \frac{1 - 3\cos^2\theta}{(2a)^3}. \tag{2B.2}$$

(2) Based on Eq. (2B. 2) for the configuration (a) in Fig. 8 – 10 we obtain

$$U_a = \frac{1}{4\pi\varepsilon_0} \frac{1-3}{(2a)^3} p^2 = -\frac{1}{4\pi\varepsilon_0} \frac{p^2}{4a^3}. \tag{2B.3}$$

For configuration (b),

$$U_b = \frac{1}{4\pi\varepsilon_0} \frac{1-0}{(2a)^3} p^2 = \frac{1}{4\pi\varepsilon_0} \frac{p^2}{8a^3}. \tag{2B.4}$$

For configuration (c),

$$U_c = \frac{1}{4\pi\varepsilon_0} \frac{1-3\cos^2\frac{\pi}{4}}{(2a)^3} p^2 = -\frac{1}{4\pi\varepsilon_0} \frac{p^2}{16a^3}. \tag{2B.5}$$

(3) Comparison between (2B. 3), (2B. 4) and (2B. 5) shows that configuration (a) has the lowest energy, corresponding to the ground state of the system.

2. With the similar approach to question 1 the interaction energies for the three different configurations can also be calculated.

in Fig. 8 – 11, for configuration (a),

$$U_d = \frac{1}{4\pi\varepsilon_0} \left(-\frac{p^2}{4a^3} + 2 \times \frac{p^2}{32a^3} \right) = -\frac{1}{4\pi\varepsilon_0} \frac{3p^2}{16a^3}. \tag{2B.6}$$

For configuration (b),

$$U_e = \frac{1}{4\pi\varepsilon_0} \left(-\frac{p^2}{4a^3} \times 2 - \frac{p^2}{32a^3} \right) = -\frac{1}{4\pi\varepsilon_0} \frac{17p^2}{32a^3}. \tag{2B.7}$$

For configuration (c),

$$U_f = \frac{1}{4\pi\varepsilon_0} \left(\frac{p^2}{8a^3} \times 2 + \frac{p^2}{64a^3} \right) = \frac{1}{4\pi\varepsilon_0} \frac{17p^2}{64a^3}. \tag{2B.8}$$

Comparison shows that configuration (b) of the lowest energy is most stable, while configuration (c) of the highest energy is most unstable.

Problem 3

3A. Average Contribution of Each Electron to Specific Heat of Free Electron Gas at Constant Volume

1. According to the classical physics the conduction electrons in metals constitute free electron gas like an ideal gas. In thermal equilibrium their average energy relates to temperature, therefore they contribute to the specific heat. The average contribution of each electron to the specific heat of free electron gas at constant volume is defined as

$$c_V = \frac{\mathrm{d}\,\overline{E}}{\mathrm{d}T}, \tag{1}$$

where $\overline{E}$ is the average energy of each electron. However the value of the specific heat at constant volume is a constant, independent of temperature. Please calculate $\overline{E}$ and the average contribution of each electron to the specific heat at constant volume c_V.

2. Experimentally it has been shown that the specific heat of the conduction electrons at constant volume in metals depends on temperature, and the experimental value at room temperature is about two orders of magnitude lower than its classical counterpart. This is because the electrons obey the quantum statistics rather than classical statistics. According to the quantum theory, for a metallic material the density of states of conduction electrons (the number of electronic states per unit volume and per unit energy) is proportional to the square root of electron energy E, then the number of states within energy range $\mathrm{d}E$ for a metal of volume V can be written as

$$\mathrm{d}S = CVE^{\frac{1}{2}}\,\mathrm{d}E, \tag{2}$$

where C is the normalization constant, determined by the total number of electrons of the system.

The probability that the state of energy E is occupied by electron is

$$f(E) = \frac{1}{1 + \exp\left(\dfrac{E - E_F}{k_B T}\right)}, \tag{3}$$

where $k_B = 1.381 \times 10^{-23}$ J $\cdot$ K^{-1} is the Boltzmann constant and T is the absolute temperature, while E_F is called Fermi level. Usually at room temperature E_F is about several eVs for metallic materials (1 eV $= 1.602 \times 10^{-19}$ J). $f(E)$ is called Fermi distribution function shown schematically in Fig. 8 - 16.

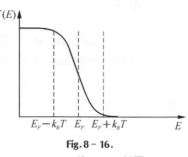

Fig. 8 - 16.

(1) Please calculate c_V at room temperature according to $f(E)$.

(2) Please give a reasonable explanation for the deviation of the classical result from that of quantum theory.

Note: In your calculation the variation of the Fermi level E_F with temperature could be neglected, i.e. assume $E_F = E_F^0$, E_F^0 is the Fermi level at 0 K. Meanwhile the Fermi distribution function could be simplified as a linearly descending function within an energy range of $2k_BT$ around E_F, otherwise either 0 or 1, i.e.

$$f(E) = \begin{cases} 1, & E < E_F - k_BT, \\ \text{linearly descending function}, & E_F - k_BT < E < E_F + k_BT, \\ 0, & E > E_F + k_BT. \end{cases}$$

At room temperature $k_BT \ll E_F$, therefore calculation can be simplified accordingly. Meanwhile, the total number of electrons can be calculated at 0 K.

3B. The Inverse Compton Scattering

By collision with relativistic high energy electron, a photon can get energy from the high energy electron, i.e. the energy and frequency of the photon increases because of the collision. This is so-called inverse Compton scattering. Such kind of phenomenon is of great importance in astrophysics, for example, it provides an important mechanism for producing X rays and γ rays in space.

1. A high energy electron of total energy E (its kinetic energy is higher than static energy) and a low energy photon (its energy is less than the static energy of an electron) of frequency ν move in opposite directions, and

collide with each other. As shown in the figure below, the collision scatters the photon, making the scattered photon move along the direction which makes an angle θ with its original incident direction (the scattered electron is not shown in the figure). Calculate the energy of the scattered photon, expressed in terms of E, ν, θ and static energy E_0 of the electron. Show the value of θ, at which the scattered photon has the maximum energy, and the value of this maximum energy.

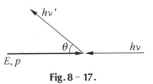

Fig. 8 – 17.

2. Assume that the energy E of the incident electron is much higher than its static energy E_0, which can be shown as $E = \gamma E_0$, $\gamma \gg 1$, and that the energy of the incident photon is much less than E_0/γ, show the approximate expression of the maximum energy of the scattered photon. Taking $\gamma = 200$ and the wavelength of the incident visible light photon $\lambda = 500$ nm, calculate the approximate maximum energy and the corresponding wavelength of the scattered photon.

Parameters. Static energy of the electron $E_0 = 0.511$ MeV, Planck constant $h = 6.63 \times 10^{-34}$ J $\cdot$ s, and $hc = 1.24 \times 10^3$ eV $\cdot$ nm, where c is the light speed in the vacuum.

3. (1) A relativistic high energy electron of total energy E and a photon move in opposite directions and collide with each other. Show the energy of the incident photon, of which the photon can gain the maximum energy from the incident electron. Calculate the energy of the scattered photon in this case.

(2) A relativistic high energy electron of total energy E and a photon, moving in perpendicular directions respectively, collide with each other. Show the energy of the incident photon, of which the photon can gain the maximum energy from the incident electron. Calculate the energy of the scattered photon in this case.

Solution

3A. Average specific heat of each free electron at constant volume

(1) Each free electron has 3 degrees of freedom. According to the

equipartition of energy theorem, at temperature T its average energy $\overline{E}$ equals to $\dfrac{3}{2}k_BT$, therefore the average specific heat c_V equals to

$$c_V = \frac{\mathrm{d}\,\overline{E}}{\mathrm{d}T} = \frac{3}{2}k_B.$$

(2) Let U be the total energy of the electron gas, then

$$U = \int_0^S Ef(E)\,\mathrm{d}S,$$

where S is the total number of the electronic states, E the electron energy. Substitution of (1) for $\mathrm{d}S$ in the above expression gives

$$U = CV\int_0^\infty E^{\frac{3}{2}} f(E)\,\mathrm{d}E = CVI,$$

where I represents the integral

$$I = \int_0^\infty E^{\frac{3}{2}} f(E)\,\mathrm{d}E.$$

Usually at room temperature $k_BT \ll E_F$. Therefore, with the simplified $f(E)$

$$f(E) = \begin{cases} 1, & E < E_F - k_BT, \\[2mm] -\dfrac{E-(E_F+k_BT)}{2k_BT}, & E_F - k_BT < E < E_F + k_BT, \\[2mm] 0, & E > E_F + k_BT. \end{cases}$$

I can be simplified as

$$\begin{aligned} I = {} & \frac{2}{5}E_F^{\frac{3}{2}}\left(1 - k_B\frac{T}{E_F}\right)^{\frac{5}{2}} + \frac{E_F + k_BT}{5k_BT}E_F^{\frac{3}{2}}\left[\left(1 + k_B\frac{T}{E_F}\right)^{\frac{5}{2}}\right. \\[2mm] & \left. - \left(1 - k_B\frac{T}{E_F}\right)^{\frac{5}{2}}\right] - \frac{1}{7k_BT}E_F^{\frac{7}{2}}\left[\left(1 + k_B\frac{T}{E_F}\right)^{\frac{7}{2}}\right. \\[2mm] & \left. - \left(1 - k_B\frac{T}{E_F}\right)^{\frac{7}{2}}\right] \\[2mm] \approx {} & E_f^{\frac{5}{2}}\left[\frac{2}{5} + \frac{3}{4}\left(k_B\frac{T}{E_F}\right)^2\right]. \end{aligned}$$

Therefore

$$U = CVE_F^{\frac{5}{2}}\left[\frac{2}{5} + \frac{3}{4}\left(k_B \frac{T}{E_F}\right)^2\right].$$

However the total electron number

$$N = CV\int_0^{E_F^0} E^{\frac{1}{2}}\, dE = \frac{2}{3}CV(E_F^0)^{\frac{3}{2}},$$

$$CV = \frac{3}{2}N(E_F^0)^{-\frac{3}{2}},$$

where E_F^0 is the Fermi level at 0 K, leading to

$$U = \frac{3}{2}N(E_F^0)^{-\frac{3}{2}} E_F^{\frac{5}{2}}\left[\frac{2}{5} + \frac{3}{4}\left(k_B \frac{T}{E_F}\right)^2\right].$$

Taking $E_F \approx E_F^0$, and $U = N\overline{E}$,

$$\overline{E} \approx \frac{3}{2}E_F\left[\frac{2}{5} + \frac{3}{4}\left(k_B \frac{T}{E_F}\right)^2\right]$$

$$c_v = \frac{\partial \overline{E}}{\partial T} = \frac{9}{4}k_B \frac{k_B T}{E_F} \ll \frac{3}{2}k_B.$$

(3) Because at room temperature $k_B T = 0.026$ eV while Fermi level of metals at room temperature is generally of several eVs, it can be seen from the above expression that according to the quantum theory the calculated average specific heat of each free electron at constant volume is two orders of magnitude lower than that of the classical theory. The reason is that with the temperature increase the energy of those electrons whose energy is far below Fermi level (several times of $k_B T$ less than E_F) does not change obviously, only those minor electrons of energy near E_F contribute to the specific heat, resulting in a much less value of the average specific heat.

3B. The Inverse Compton Scattering

1. Let p and E denote the momentum and energy of the incident electron, p' and E' the momentum and energy of the scattered electron, and $h\nu$ and $h\nu'$ the energies of the incident and scattered photon respectively. For this scattering process (see Fig. 8 – 18) energy conservation reads

$$h\nu + E = h\nu' + E', \tag{3B.1}$$

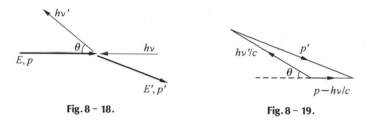

Fig. 8 – 18.   **Fig. 8 – 19.**

while the momentum conservation can be shown as (see Fig. 8 – 19)

$$(p'c)^2 = (h\nu')^2 + (pc - h\nu)^2 + 2h\nu'(pc - h\nu)\cos\theta. \qquad (3B.2)$$

Eqs. (3B.1) and (3B.2), combined with the energy-momentum relations

$$E^2 = (pc)^2 + E_0^2 \qquad (3B.3)$$

and

$$E'^2 = (p'c)^2 + E_0^2 \qquad (3B.4)$$

lead to

$$h\nu' = \frac{E + pc}{E + h\nu + (pc - h\nu)\cos\theta} h\nu$$

$$= \frac{E + \sqrt{E^2 - E_0^2}}{E + h\nu + (\sqrt{E^2 - E_0^2} - h\nu)\cos\theta} h\nu. \qquad (3B.5)$$

We have assumed that the kinetic energy of the incident electron is higher than its static energy, and the energy of the incident photon $h\nu$ is less than E_0, so that $\sqrt{E^2 - E_0^2} > h\nu$. Therefore from Eq. (3B.5), it can be easily seen that $\theta = \pi$ results in the maximum of $h\nu'$, and the maximum $h\nu'$ is

$$(h\nu')_{max} = \frac{E + \sqrt{E^2 - E_0^2}}{E + 2h\nu - \sqrt{E^2 - E_0^2}} h\nu. \qquad (3B.6)$$

2. Substitution of $E = \gamma E_0$ into Eq. (3B.6) yields

$$(h\nu')_{max} = \frac{\gamma E_0 + \sqrt{\gamma^2 - 1}E_0}{\gamma E_0 - \sqrt{\gamma^2 - 1}E_0 + 2h\nu} h\nu$$

$$= \frac{\gamma + \sqrt{\gamma^2 - 1}}{\gamma - \sqrt{\gamma^2 - 1} + \dfrac{2h\nu}{E_0}} h\nu. \qquad (3B.7)$$

Due to $\gamma \gg 1$, $\sqrt{\gamma^2 - 1} \approx \gamma\left(1 - \dfrac{1}{2\gamma^2}\right) = \gamma - \dfrac{1}{2\gamma}$, and $h\nu/E_0 \ll 1/\gamma$, then we have

$$(h\nu')_{max} \approx \frac{\gamma + \gamma - \dfrac{1}{2\gamma}}{\gamma - \gamma + \dfrac{1}{2\gamma} + \dfrac{2h\nu}{E_0}} h\nu \approx 4\gamma^2 h\nu. \tag{3B.8}$$

In the case of $\gamma = 200$ and the wavelength of the incident photon $\lambda = 500$ nm,

$$h\nu = \frac{hc}{\lambda} = \frac{1.24 \times 10^3}{500} = 2.48 \text{ eV},$$

$$\frac{h\nu}{E_0} = \frac{2.48}{0.511 \times 10^6} = 4.85 \times 10^{-6} \ll \frac{1}{\gamma} = \frac{1}{200} = 5.0 \times 10^{-3},$$

satisfying expression (3B.8). Therefore the maximum energy of the scattered photon

$$(h\nu')_{max} \approx 4 \times 200^2 \, h\nu = 1.6 \times 10^5 \times 2.48 = 3.97 \times 10^5 \text{ eV}$$
$$\approx 4.0 \times 10^5 \text{ eV} = 0.40 \text{ MeV}$$

corresponding to a wavelength $\lambda' = \dfrac{hc}{h\nu'} = \dfrac{1.24 \times 10^3}{4.0 \times 10^5} = 3.1 \times 10^{-3}$ nm.

3. (1) It is obvious that if the incident electron gives its total kinetic energy to the photon, the photon gains the maximum energy from the incident electron through the scattering process, namely the electron should become at rest after the collision. In this case, we have (see Fig. 8 – 20)

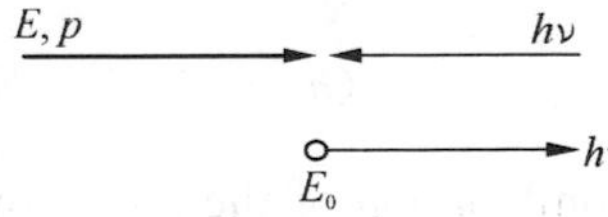

Fig. 8 – 20.

$$h\nu + E = h\nu' + E_0 \text{ (Conservation of energy)} \tag{3B.9}$$

$$p - \frac{h\nu}{c} = \frac{h\nu'}{c} \text{ (Conservation of momentum)}$$

or
$$pc - h\nu = h\nu'. \tag{3B.10}$$

Subtracting (3B.10) from (3B.9) leads to the energy of the incident photon

$$h\nu = \frac{1}{2}(E_0 - E + pc)$$

$$= \frac{1}{2} (E_0 - E + \sqrt{E^2 - E_0^2}). \qquad (3B.11)$$

In above equation the energy-momentum relation

$$(pc)^2 = E^2 - E_0^2 \qquad (3B.12)$$

has been taken into account. Therefore from Eq. (3B. 9) we obtain the energy of the scattered photon

$$hv' = hv + E - E_0$$

$$= \frac{1}{2} (E - E_0 + \sqrt{E^2 - E_0^2}). \qquad (3B.13)$$

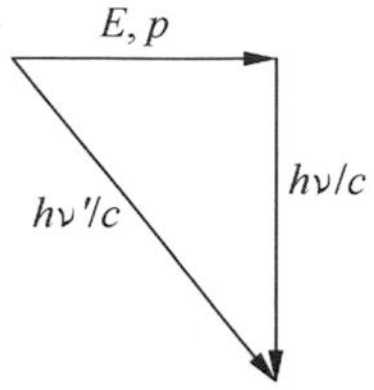

(2) Similar to question 3. (1), now we have (see Fig. 8 – 21)

Fig. 8 – 21.

$$hv + E = hv' + E_0, \quad \text{(Conservation of energy)} \qquad (3B.9)$$

$$p^2 + \left(\frac{hv}{c}\right)^2 = \left(\frac{hv'}{c}\right)^2 \text{(Conservation of momentum)}.$$

That is,

$$(pc)^2 + (hv)^2 = (hv')^2. \qquad (3B.14)$$

Substitution of Eq. (3B. 12) into Eq. (3B. 14) yields

$$E^2 - E_0^2 + (hv)^2 = (hv')^2.$$

On the other hand, square of Eq. (3B.9) results in

$$(hv')^2 = E^2 + E_0^2 + (hv)^2 + 2Ehv - 2EE_0 - 2E_0 hv.$$

Combination of the above two equations leads to

$$E^2 + E_0^2 + (hv)^2 + 2Ehv - 2EE_0 - 2E_0 hv = E^2 - E_0^2 + (hv)^2.$$

That is,

$$2(E - E_0)hv = 2E_0 (E - E_0).$$

Then, we obtain the energy of the incident photon

$$hv = E_0. \qquad (3B.15)$$

Substitution of (3B. 15) into Eq. (3B. 9) gives the energy of the scattered photon

$$hv' = hv + E - E_0 = E. \qquad (3B.16)$$

Explanatory notes about the solution of Question 3:

Question 3. (1) can also be solved as follows. According to Eq. (3B.6), the maximum energy Δ that the photon of energy $h\nu$ gains from the electron is

$$\Delta = h\nu'_{max} - h\nu = 2\frac{pch\nu - h^2\nu^2}{E + 2h\nu - pc},$$

where $pc = \sqrt{E^2 - E_0^2}$. To obtain the maximum Δ, we use the extreme condition

$$\frac{d\left(\frac{\Delta}{2}\right)}{d(h\nu)} = \frac{(pc - 2h\nu)(E + 2h\nu - pc) - 2(pch\nu - h^2\nu^2)}{(E + 2h\nu - pc)^2} = 0.$$

Let the numerator equal to zero, a quadratic equation results:

$$2(h\nu)^2 + 2(E - pc)h\nu - (Epc - p^2c^2) = 0.$$

Its two roots can be shown as

$$h\nu = \frac{1}{2}\left[-(E - pc) \pm \sqrt{E^2 - p^2c^2}\right] = \frac{1}{2}(-E + pc \pm E_0).$$

Since the negative sign leads to a meaningless negative $h\nu$, we have

$$h\nu = \frac{1}{2}(\sqrt{E^2 - E_0^2} - E + E_0),$$

where $pc = \sqrt{E^2 - E_0^2}$ has been taken into account. This result is just the same as Eq. (3B.11). The expression for $h\nu'$ is then the same as Eq. (3B. 13).

Question 3. (2) can also be solved as follows

For the sake of simplification, it is assumed that the scattered photon and electron move in the same plane which the incident photon and electron moved in. Meanwhile the angles which the directions of the scattered photon and electron make with the direction of the incident electron are denoted by Ψ and φ respectively (see the figure). Then, we have

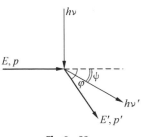

Fig. 8 – 22.

$$E + h\nu = E' + h\nu', \text{ (Conservation of energy)} \qquad (3B.1')$$

$$p = \frac{h\nu'}{c}\cos\Psi + p'\cos\varphi,$$

or

$$pc - h\nu'\cos\Psi = p'c\cos\varphi, \text{ (Conservation of horizontal momentum)}$$
$$(3B.2')$$

and
$$\frac{h\nu}{c} = \frac{h\nu'}{c}\sin\Psi + p'\sin\varphi,$$

or

$$h\nu - h\nu'\sin\Psi = p'c\sin\varphi. \text{ (Conservation of vertical momentum)}$$
$$(3B.3')$$

$(3B.2')^2 + (3B.3')^2$ leads to

$$p'^2c^2 = p^2c^2 + (h\nu')^2\cos^2\Psi - 2pch\nu'\cos\Psi + (h\nu)^2$$
$$+ (h\nu')^2\sin^2\Psi - 2h\nu h\nu'\sin\Psi. \qquad (3B.4')$$

Square of Eq. $(3B.1')$ results in

$$E'^2 = E^2 + (h\nu)^2 + (h\nu')^2 + 2Eh\nu - 2Eh\nu' - 2h\nu h\nu'.$$

Substitution of the energy-momentum relation

$$E'^2 = E_0^2 + p'^2c^2, \quad E^2 = E_0^2 + p^2c^2$$

into the above equation of energy conservation leads to

$$p'^2c^2 = p^2c^2 + (h\nu)^2 + (h\nu')^2 + 2Eh\nu - 2Eh\nu' - 2h\nu h\nu'. \quad (3B.5')$$

Comparison between Eq. $(3B.4')$ and Eq. $(3B.5')$ yields the energy of the scattered photon

$$h\nu' = \frac{Eh\nu}{E + h\nu - (pc\cos\Psi + h\nu\sin\Psi)}. \qquad (3B.6')$$

From $(3B.6')$ it can be seen that if $\Psi = \cos^{-1}\dfrac{pc}{\sqrt{p^2c^2 + h^2\nu^2}}$, the energy of the scattered photon reaches the maximum

$$h\nu'_{max} = \frac{Eh\nu}{E + h\nu - \sqrt{p^2c^2 + h^2\nu^2}}. \qquad (3B.7')$$

The energy that the photon gets from the electron is

$$\Delta = h\nu'_{max} - h\nu = \frac{\sqrt{p^2 c^2 + h^2 \nu^2}\, h\nu - (h\nu)^2}{E + h\nu - \sqrt{p^2 c^2 + h^2 \nu^2}}. \tag{3B.8'}$$

The extreme condition for Δ is

$$\frac{d\Delta}{d(h\nu)} = \frac{A}{(E + h\nu - a)^2} = 0, \tag{3B.9'}$$

where

$$\sqrt{p^2 c^2 + h^2 \nu^2} = a$$

and

$$A = \left[a + \frac{(h\nu)^2}{a} - 2h\nu \right](E + h\nu - a) - (h\nu a - h^2 \nu^2)\left(1 - \frac{h\nu}{a}\right).$$

$A = 0$ results in

$$Ea + h\nu a - p^2 c^2 - (h\nu)^2 + \frac{E(h\nu)^2}{a} + \frac{(h\nu)^3}{a} - (h\nu)^2 - 2h\nu E$$

$$- 2(h\nu)^2 + 2h\nu a = h\nu a - (h\nu)^2 - (h\nu)^2 + \frac{(h\nu)^3}{a}.$$

Simplifying this equation leads to

$$Ea - p^2 c^2 + \frac{E(h\nu)^2}{a} - 2h\nu E - 2(h\nu)^2 + 2h\nu a = 0,$$

i. e.

$$(E + 2h\nu)a + \frac{E(h\nu)^2}{a} = p^2 c^2 + 2h\nu E + 2(h\nu)^2.$$

Squaring both of the two sides of this equation yields

$$(E + 2h\nu)^2 a^2 + \frac{E^2 (h\nu)^4}{a^2} + 2(E + 2h\nu)E(h\nu)^2$$

$$= (pc)^4 + 4(h\nu)^2 E^2 + 4(h\nu)^4 + 4(pc)^2 h\nu E$$

$$+ 4(pc)^2 (h\nu)^2 + 8(h\nu)^3 E.$$

Substitution of $\sqrt{p^2 c^2 + h^2 \nu^2} = a$ into the above equation and making some simplifications yield

$$E^2 p^2 c^2 + \frac{E^2 (h\nu)^4}{p^2 c^2 + h^2 \nu^2} = p^4 c^4 + h^2 \nu^2 E^2,$$

that is,

$$E^2 (pc)^4 + E^2 (pc)^2 (h\nu)^2 + E^2 (h\nu)^4$$
$$= (pc)^6 + (pc)^4 (h\nu)^2 + E^2 (pc)^2 (h\nu)^2 + (h\nu)^4 E^2.$$

After some simplifications we obtain

$$(pc)^4 (h\nu)^2 = (pc)^4 (E^2 - p^2 c^2) = (pc)^4 E_0^2,$$

which yields
$$h\nu = E_0. \tag{3B.10'}$$

Substitution of $(3B.10')$ into Eq. $(3B.7')$ leads to

$$h\nu'_{max} = \frac{EE_0}{E + E_0 - E} = E. \tag{3B.11'}$$

The results $(3B.10')$ and $(3B.11')$ are just the same as Eqs. $(3B.15)$ and $(3B.16)$ in the former solution.

Experimental Competition

April 25,2007

Before attempting to assemble your apparatus,
read the problem text completely!

Problem 1

Using the Interference Method to Measure the Thermal Expansion Coefficient and Temperature Coefficient of Refractive index of Glass

(1) Instructions

Optical instruments are often used at high or low temperatures. When optical instruments are used at different temperatures, the thermal properties of the materials, of which the optical elements were made of, including thermal expansion and the variation of refractive index with temperature, will directly affect their optical properties. Two parameters, i.e. the linear thermal expansion coefficient β and the temperature coefficient of refractive index γ, are defined as $\beta = \dfrac{1}{L} \dfrac{dL}{dT}$ and $\gamma = \dfrac{dn}{dT}$ respectively to describe these properties, where L stands for the length of the material, T the temperature, and n the refractive index. The purpose of the present experiment is to measure β and γ of a given glass material.

(2) Experimental apparatus, devices, and materials

1. **Sample.** The cylinder-like sample used in present experiment is made of uniform and isotropic glass, as shown in Fig. 8 – 23, where A represents a glass cylinder with a segmental part parallel to its axis cut away, its top and bottom surface are approximately parallel to each other. B and B' are two circular plates made of the same glass material, from each of which a

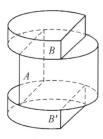

Fig. 8 – 23.

segmental part parallel to the axis was also cut away. The top and bottom surface of each glass plate are not parallel to each other. A, B, and B' are glued together as a whole, as shown in Fig. 8 – 23. The refractive index of the glue is as the same as that of the glass, and its thickness can be neglected.

2. Heater. The heater used in this experiment is schematically shown in Fig. 8 – 24. A knob on the right of the electric oven is used to adjust the temperature of the electric oven. A big aluminum cylinder is bored a co-axis pipe-like sample cavity in the middle of it. The experimental sample can be put at the bottom of this cavity. Besides, there is a small aluminum cylinder, through which two tube-like holes of different radius were bored. The small hole allows light to pass through, while a probe of a thermometer can be inserted in the big hole. If you want to heat the sample, you should first slip the sample carefully into the cavity of the big aluminum cylinder before you heat it (in doing so, the big aluminum cylinder should be inclined somewhat to avoid cracking the sample). Next, put the small aluminum cylinder onto the sample already located in the cavity. Then, put the whole big aluminum cylinder including the sample and the small cylinder inside into the steel cup on the electric oven. (For heating, you should not put any water in the steel cup, and you should not take away the steel cup but put the whole aluminum cylinder directly on the oven, either.)

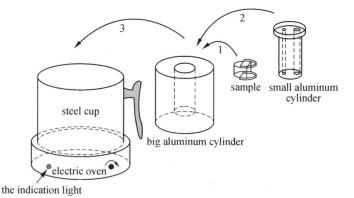

Fig. 8 – 24.

3. Light source holder. As shown in Fig. 8 – 25, a support consisting of a vertical post and a base trestle is designed to hold the laser light source. The post and the base trestle are fixed together at C and two tunable adjusting screws A and B are attached to the trestle. A He – Ne laser and its power supply are held at the upper part of the post, as shown in Fig. 8 – 25. Just below the laser, a slant bracket is attached to the power supply, on which is placed an aluminum plate with a hole through it. A piece of graph paper is attached to the aluminum plate, which can be used as an observation screen.

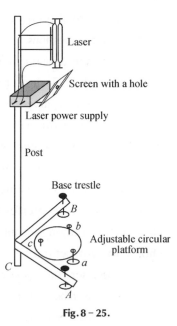

Fig. 8 – 25.

4. Sample platform. A circular platform with three adjusting screws a, b, and c is designed to hold the heating oven or the big aluminum cylinder including the sample and the small aluminum cylinder. The platform is placed on the experimental table, close to the base trestle, right below the laser light source, as shown in Fig. 8 – 25.

5. A digital thermometer.

6. A straight ruler.

7. A basin(containing the cooling water).

8. A piece of towel.

9. A piece of graph paper.

10. A calculator.

11. A pen and a pencil.

Attention:

1. Never look at the laser light source along the direction opposite to the incident direction of the light. Otherwise, the laser light might hurt your eyes.

2. Never touch the optical surface of the sample. Take the sample gently to avoid any damage.

(3) Experimental content

1. Answer Questions

1. 1   When a beam of white natural light coming from a lamp and passing through a piece of red transparent paper strikes on a thick glass slab of a thickness about 2 cm, the beams reflected from the top and bottom surfaces of the slab meet at the observation screen, resulting in a light spot on the screen without interference fringes, as shown in Fig. 8 – 26(a). However, when a laser beam strikes on the same glass slab, the reflected beams meet at the same observation screen, resulting in a light spot with interference fringes, as shown in Fig. 8 – 26 (b). What is the reason accounting for these two different phenomena? (choose the correct one)

A. The laser beam is stronger than the red beam.

B. The laser beam is more collimated than the red beam.

C. The laser beam is more coherent than the red beam.

D. The wavelength of the laser beam is shorter than that of the red beam.

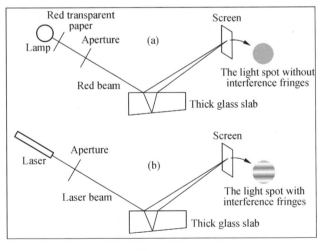

Fig. 8 – 26.

1. 2(a)   As shown in Fig. 8 – 27 if a laser light beam strikes approximately normally on the region a, b, and c of the sample respectively, how many

main light spots of the reflection light will be observed
(need not consider multiple reflections)? Will the profiles
of these reflected optical spots be inevitably the same as
that of the incident light?

1. 2(b)　If the profile of some spots of the reflected
light is different from that of the incident light? Please
give a reason to account for.

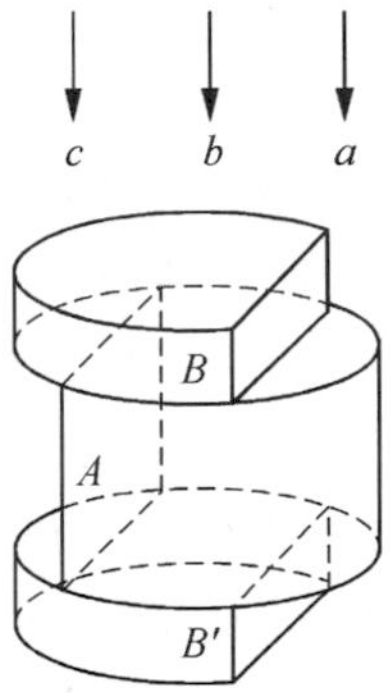

2. Experiment

Measurement of β and γ of the glass.

Fig. 8 – 27.

With the wavelength of the laser light $\lambda = 632.8$ nm,
the height of the glass cylinder A $L = 10.12 \pm 0.05$ mm, and the average
refractive index of the glass corresponding to this given wavelength and the
given temperature range of measurement $n = 1.515$, measure the thermal
expansion coefficient β and the temperature coefficient of the refractive
index γ for the glass sample over the temperature range from 40℃ to 90℃.
(Within this temperature range β and γ may be taken as constant.)

2. 1　Design the Experiment, draw the experimental ray diagrams and
derive the formulae relevant to the measurement.

2. 2　Carry out the experiment and record the measured data of the
thermal expansion coefficient β and γ.

2. 3　Calculate β and γ of the given glass material and estimate their
uncertainties.

2. 4　Write the final experimental results.

Attention:

1. When put the sample into the sample cavity of the big aluminum
cylinder, in order to avoid cracking the sample, incline somewhat the big
aluminum cylinder, and make the sample slip towards the bottom of the
sample cavity carefully and slowly.

2. During the process of experiment, the temperature can be
increased continuously. In order to guarantee enough time you better to
measure your data when the temperature is increasing. To heat the
sample, the knob of the electric oven may be first turned to its maximum.

When the temperature approaches to near 90℃ (around 85℃), turn the knob to the minimum immediately to stop heating. During heating, the indication light on the left of the oven might be off and then on again. This indicates that the oven is controlling the temperature itself automatically. Do not care about it.

3. After the sample in the cavity of the big aluminum cylinder is heated enough, its natural cooling down will be very slow. To get a rapid cooling, you may put the heated steel cup containing the big aluminum cylinder with the small aluminum cylinder inside into the water-filled basin. After a while of cooling (at least 5 minutes), carefully use the towel to wrap the big aluminum cylinder and put it directly into the cooling water to speed up the cooling process. Be careful to avoid burning your hand. After cooling, use the towel to dry the big aluminum cylinder, and put it back to the steel cup, then you can heat it again. In order to avoid short circuit, never pour any water into the electric oven.

4. When you finished the experiment, turn off the electric oven immediately to avoid overheating.

Warning: Be careful in your experiments! If your sample or instrument is broken, you may have difficulties to continue your experiments, since we do not have enough backups!

Solution

1. Questions

1.1　C

1.2(a)　Fill in the blanks

Region	Number of reflected light spots observed	Must the profile of the reflected light spots be the same as that of the incident light? If "Yes", fill in "Y"; if "No", fill in "N"				
		1st (if any)	2nd (if any)	3rd (if any)	4th (if any)	...
a	1	N				

Cont.

Region	Number of reflected light spots observed	Must the profile of the reflected light spots be the same as that of the incident light? If "Yes", fill in "Y"; if "No", fill in "N"				
		1st (if any)	2nd (if any)	3rd (if any)	4th (if any)	...
b	2	Y	Y			
c	3	Y	N	Y		

Illustration:

Region a: Because the upper and lower surface of the glass cylinder A are approximately parallel to each other, when the laser beam arrives on the cylinder nearly perpendicular to the surface, the reflected light spots will overlap each other, causing interference fringes. Therefore the distribution of light intensity would be different from that of the incident light.

Region b: Because the refractive index of the glue is the same as that of the glass and its thickness is negligible, no light will be reflected from the interface between them. However the upper and lower surface of each glass plate are not parallel to each other, the upper surface of the upper plate will also be not parallel to the lower surface of the lower plate, therefore the two reflected beams from each of them will form two light spots and their distribution of the light intensity would surely be the same as that of the incident light.

Region c: Because the upper and lower surface of each glass plate are not parallel to each other, but the upper and lower surface of the glass cylinder are approximately parallel to each other, there must be a pair of reflected light spots from the two plates overlap each other, causing interference fringes, while the light intensity distribution of the other two reflected light spots would be the same as that of the incident light.

1. 2(b)   If you choose "No" (N), use one keyword to account for the reason: interference.

2. Experiment: Measuring β and γ

2. 1   Design the Experiment, draw the experimental ray diagrams and derive the formulae relevant to the measurement.

The experimental ray diagrams for measuring β (left) and γ (right) are shown in Fig. 8 – 28.

When the laser light is reflected from the c region of the sample as shown in Fig. 8 – 28 (left), three reflected light spots can be observed on the screen, some interference fringes appear at spot v, which are caused by the interference between the two light rays reflected from the bottom surface

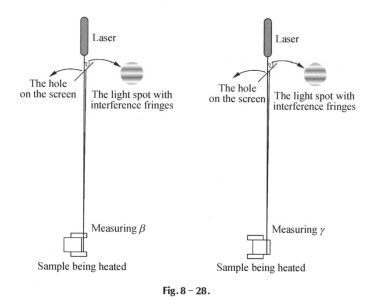

Fig. 8 – 28.

of the upper glass plate and the top surface of the lower glass plate. The difference between the optical lengths of the two light rays is $2L$. After the electric oven starts heating, assume that the temperature T has increased by ΔT, the length increment of the sample due to the thermal expansion of glass will be $\Delta L = L\beta\Delta T$, and the shift in the number of the moving interference fringes is m_1. Then, $2\Delta L = m_1\lambda$, where λ stands for the wavelength of the laser light. Thus,

$$\beta = \frac{m_1\lambda}{2L\Delta T}.$$

With the given L and λ, from the graphic relation of m_1 and T we obtain the shift in the number of the moving fringes m_1 over the temperature range from 40℃ to 90℃. Then, β can be obtained.

When the laser light is reflected from the region a as shown in Fig. 8 – 28 (right), the difference between the optical paths is $2nL$. The variation of optical path difference caused by temperature increase ΔT is

$$\Delta(2nL) = 2\left(n\frac{\Delta L}{\Delta T} + L\frac{\Delta n}{\Delta T}\right)\Delta T$$
$$= 2L(n\beta + \gamma)\Delta T.$$

Assume that at this time the shift in the number of the moving interference fringes is m_2,

$$2L(n\beta + \gamma)\Delta T = m_2\lambda,$$

i. e.

$$\gamma = \frac{m_2\lambda}{2L\Delta T} - n\beta = \left(\frac{m_2}{m_1} - n\right) \times \beta.$$

From the graphic relation of $m_2 \sim T$ we obtain the shift in the number of the moving interference fringes m_2 over the temperature range from 40℃ to 90℃. Thus, γ can be obtained.

2. 2　(1) Data recorded during the measurement of the thermal expansion coefficient β

Measured Relation of m_1 and T:

m_1 (fringes)	1	2	3	4	5	6	7	8
T(℃)	30. 0	35. 4	40. 6	46. 1	50. 6	54. 4	58. 6	63. 1
m_1 (fringes)	9	10	11	12	13	14	15	
T(℃)	67. 6	72. 2	75. 8	79. 8	83. 8	87. 4	90. 9	

(2) Data recorded during the measurement of the temperature coefficient of the refraction index γ

Measured relation of m_2 and T:

m_2 (fringes)	1	2	3	4	5	6	7	8	9	10
T (℃)	25. 4	27. 0	28. 9	31. 0	33. 4	35. 3	37. 6	40. 0	42. 2	44. 4
m_2 (fringes)	11	12	13	14	15	16	17	18	19	20
T(℃)	46. 6	48. 9	51. 6	54. 0	56. 2	58. 6	61. 2	64. 0	66. 4	69. 0
m_2 (fringes)	21	22	23	24	25	26	27	28	29	30
T(℃)	72. 4	75. 0	77. 6	79. 8	82. 4	84. 4	86. 4	88. 2	90. 2	

2. 3　Get the thermal expansion coefficient β and the temperature coefficient of refractive index γ and estimate their uncertainties.

(1) Draw the graphic relation of $m_1 \sim T$ and $m_2 \sim T$.

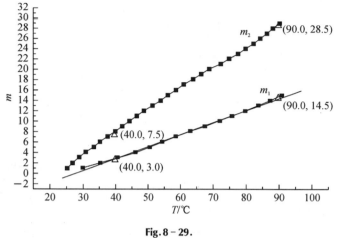

Fig. 8 – 29.

(2) Calculate β.

With the parameters: $L = 10.12 \pm 0.05$ mm, $\lambda = 632.8$ nm, $\Delta T = 50.0$℃, and $m_1 = 11.5$ (over temperature from 40℃ to 90℃) obtained from Fig. 8 – 26, we get

$$\beta = \frac{m_1 \lambda}{2L\Delta T} = 7.19 \times 10^{-6}\,℃^{-1}.$$

(3) Estimate the uncertainty of β.

With $u(L) = 0.05$ mm, $u(\Delta T) = 0.2$℃, and estimation of $u(m_1) = 0.2$, we get

$$\left(\frac{u(\beta)}{\beta}\right)^2 = \left(\frac{u(m_1)}{m_1}\right)^2 + \left(\frac{u(L)}{L}\right)^2 + \left(\frac{u(\Delta T)}{\Delta T}\right)^2$$

$$= \left(\frac{0.2}{11.5}\right)^2 + \left(\frac{0.05}{10.12}\right)^2 + \left(\frac{0.2}{50}\right)^2$$

$$= 3.4 \times 10^{-4},$$

and
$$u(\beta) = 0.13 \times 10^{-6}\,℃^{-1}.$$

(4) Calculate γ.

With $n = 1.515$ and $m_1 = 11.5$ it can be obtained from the graphic relation of $m_2 \sim T$ that $m_2 = 21.0$ over temperatures from 40℃ to 90℃.

Therefore from the measured $\beta = (7.19 \pm 0.13) \times 10^{-6} °C^{-1}$ and $\gamma = \left(\dfrac{m_2}{m_1} - n\right)\beta$,

we obtain $\gamma = 2.24 \times 10^{-6} °C^{-1}$.

(5) Estimate the uncertainty of γ.

With obtained $u(\beta) = 0.13 \times 10^{-6} °C^{-1}$ and estimation $u(m_1) = u(m_2) = 0.2$,

$$\frac{u(\gamma)}{\gamma} = \sqrt{\left(\frac{u(m_2) + nu(m_1)}{m_2 - nm_1}\right)^2 + \left(\frac{u(\beta)}{\beta}\right)^2 + \left(\frac{u(m_1)}{m_1}\right)^2} = 0.13$$

$$u(\gamma) = 0.30 \times 10^{-6} °C^{-1}.$$

2.4 Experimental results.

The thermal expansion coefficient of the sample glass material is

$$\beta = (7.19 \pm 0.13) \times 10^{-6} °C^{-1}.$$

The temperature coefficient of the refractive index of the sample glass material is

$$\gamma = (2.24 \pm 0.30) \times 10^{-6} °C^{-1}.$$

Panel Instruction

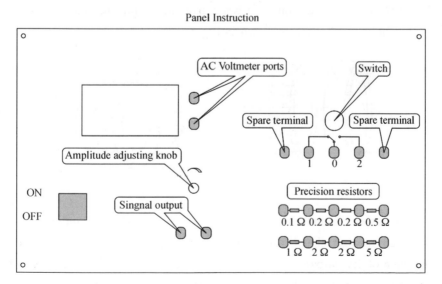

Fig. 8 - 30.

Note: The AC voltmeter black port is connected with two spare terminals.

Problem 2
Measurement of Liquid Electric Conductivity

1. Experimental instructions

In the apparatus of present experiment to measure the conductivity of liquid (i. e. water with salt), the sensor deals with ac signal without any contact potential involved to interfere with the desired experimental results. Meanwhile, since the sensor (detective winding) does not directly touch the liquid to be measured, no chemical reaction would happen during the experiments to damage any part of the apparatus. Therefore it can be used repeatedly for a long time.

As shown in Fig. 8 – 31, the sensor designed for measuring the conductivity of liquid consists of two circular loops with the same radius, made of soft-iron-based alloy. Each loop is wound with winding. The numbers of circles of the two windings are equal to each other. The two alloy loops are aligned along the same axis and connected closely as one airproof hollow cylinder, as shown in Fig. 8 – 32.

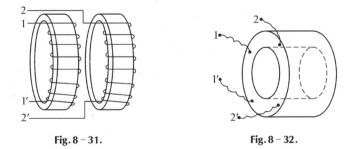

Fig. 8 – 31.   Fig. 8 – 32.

The sensor is immersed in the liquid to be measured. Winding 11' is connected to sine signal generator of frequency about 2. 5 kHz. The amplitude of its output signal might drift somewhat. If the drift exceeds certain value, it should be adjusted in time to keep the output amplitude remain at certain value. Winding 22' is connected to ac voltmeter used to measure the induced signal voltage. With the measured magnitude of the signal voltage, the conductivity of the liquid can be calculated.

2. Experimental principles

The operation principle of the present experimental apparatus can be simply explained as follows. The ac sine current from the signal generator induces an ac magnetic field in loop 11'. In turn the magnetic field induces an ac current in the conducting liquid. Such induced current induces back a time-varying magnetic field in loop 22', which induces an electromotive force in the same loop 22', being the output signal of the sensor.

Neglecting the magnetic hysteresis effect, output voltage V_o is a monotonical function of input voltage V_i. When input voltage V_i and the conductivity σ of the liquid are respectively within certain range, a proportional relation holds between σ and the ratio of $\dfrac{V_o}{V_i}$:

$$\sigma = K\left(\frac{V_o}{V_i}\right), \tag{1}$$

where K is the proportionality constant.

In the present apparatus, the liquid container can contain so much liquid to be measured that the resistance of the liquid outside the cylinder-shaped sensor is negligible. Therefore the output voltage V_o of the sensor depends mainly on the "liquid within the hollow cylinder" (referred as "liquid cylinder" hereafter). Thus, it is possible to use the liquid cylinder to calculate the liquid conductivity. Resistance of the liquid cylinder is

$$R = \frac{1}{\sigma}\frac{L}{S}, \ \sigma = \frac{1}{R}\frac{L}{S}, \tag{2}$$

where L is the length of the liquid cylinder along its axis, and S is the area of its cross section. Combination of (1) and (2) leads to:

$$\frac{V_o}{V_i} = \left(\frac{1}{K}\frac{L}{S}\right)\frac{1}{R} = B\frac{1}{R}, \tag{3}$$

where $B = \dfrac{1}{K}\dfrac{L}{S}$, or alternatively $K = \dfrac{1}{B}\dfrac{L}{S}$.

With Eqs. (2) and (3) we obtain

$$\sigma = \left(\frac{1}{B}\frac{L}{S}\right)\frac{V_o}{V_i}. \tag{4}$$

Eq. (4) shows that, when using the present sensor to measure the liquid conductivity, σ is related to L (length of the hollow cylinder), S (area of its cross section), $\frac{V_o}{V_i}$, and B as well.

Remark. Essentially in the present experiment, in order to obtain the proportionality constant K and then B accurately, various kinds of liquid with known σ should be required and prepared. Obviously this is not an easy task. Therefore, for the sake of both convenience and correctness, instead of the various liquids of known σ, we use externally connected standard resistors. The two ends of the standard resistor are connected to the two ends of a conducting thread passing through the hollow cylinder of the sensor to form a resistor circuit, as shown in Fig. 8 – 33.

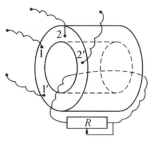

Fig. 8 – 33.

3. Experimental content

(1) Draw the experimental circuit diagram for scaling the sensor of the liquid conductivity, and complete the connection of the circuit in order to measure both the input voltage V_i (< 2.000 V) and induced output voltage V_o according to the above circuit diagram.

(2) According to the range of resistance of standard resistors: $0 \sim 9.9\ \Omega$, measure $\frac{V_o}{V_i}$ for various resistances. Record the data in the data Table designed by yourself.

Control the amplitude of V_i at any moment to make sure that its effective value is within the range of $[1.700\ \text{V},\ 1.990\ \text{V}]$ and its variation should not be higher than 0.03 V. You can also fix the input voltage at a single value within this range.

(3) (a) Take $\frac{V_o}{V_i}$ as ordinate and the reciprocal of resistance R of the standard resistor $\frac{1}{R}$ as abscissa. Draw the curve of $\frac{V_o}{V_i}$ versus $\frac{1}{R}$. The

number of measurement points should be greater than 20 within the whole output voltage range, and you are not required to add error (uncertainty) bars to the graph, but should estimate the uncertainties from the scatter points.

(3) (b) It can be seen that at some region of less induced current the curve is linear. Graph this linear part and use the graphical method to obtain the slope B of the straight part of the curve and its relative uncertainty $u(B)$ or $\dfrac{u(B)}{B}$.

(4) With the given axis length of the sensor $L = (30.500 \pm 0.025)$ mm and diameter of the liquid cylinder $d = (13.900 \pm 0.025)$mm, calculate the value of K and $u(K)$ or $\dfrac{u(K)}{K}$.

(5) Work out the conductivity of the liquid in the container and write the result. According to the uncertainties of L, d, and B, estimate the uncertainty of the conductivity. The measurement of the conductivity should be done for six times, during which the liquid should be stirred for each time.

4. Instruments and materials

(1) Sensor of liquid conductivity

The sensor has four ports of connection terminals: two terminal ports are connected to winding 11' and two terminal ports are connected to the other winding 22'.

(2) Container filled with the yet-to-be-determined liquid and stirring rod.

(3) The instrument for measuring the liquid conductivity.

On the instrument panel there are:

• Signal generator:

Two ports of connection terminals connected to the signal generator, the red one for signal output, and the black one for grounding. The amplitude of output signal can be adjusted by turning the knob.

• ac digital voltmeter.

• Inserting-type resistor box:

On the panel, there are many ports of connection terminals, between every two adjacent ports, there is a resistor with relative resistance error of 0. 001. Resistance of these resistors is 0. 1, 0. 2, 0. 5, 1, 2 and 5 Ω respectively.

● Switch 1×2(single-pole double throw).

(4) Some leads.

(5) Two pieces of graph paper(20 cm × 25 cm), calculator, recording paper, ruler, and pen.

🔑 Solution

1. Graph the experimental circuit diagram for scaling the sensor of liquid conductivity and the connection of the circuit.

2. Measure $\dfrac{V_o}{V_i}$ for different standard resistors. Record the data in the Table designed by yourself.

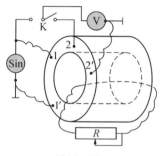

Fig. 8 – 34.

V_i(V)	V_o(V)	R(Ω)	$\dfrac{V_o}{V_i}$	$\dfrac{1}{R}$(S)
1. 95	1. 621	0. 1	0. 831	10. 000
1. 95	1. 494	0. 2	0. 766	5. 000
1. 95	1. 342	0. 3	0. 688	3. 333
1. 95	1. 216	0. 4	0. 624	2. 500
1. 95	1. 091	0. 5	0. 559	2. 000
1. 95	0. 987	0. 6	0. 506	1. 667
1. 95	0. 912	0. 7	0. 468	1. 429
1. 95	0. 837	0. 8	0. 429	1. 250
1. 95	0. 775	0. 9	0. 397	1. 111
1. 95	0. 718	1. 0	0. 368	1. 000
1. 95	0. 508	1. 5	0. 260	0. 667
1. 95	0. 396	2. 0	0. 203	0. 500

Cont.

V_i(V)	V_o(V)	$R(\Omega)$	$\dfrac{V_o}{V_i}$	$\dfrac{1}{R}$(S)
1. 95	0. 318	2. 5	0. 163	0. 400
1. 95	0. 270	3. 0	0. 138	0. 333
1. 95	0. 230	3. 5	0. 118	0. 286
1. 95	0. 201	4. 0	0. 103	0. 250
1. 95	0. 177	4. 5	0. 091	0. 222
1. 95	0. 160	5. 0	0. 082	0. 200
1. 95	0. 144	5. 5	0. 074	0. 182
1. 95	0. 132	6. 0	0. 068	0. 167
1. 95	0. 118	6. 5	0. 061	0. 154
1. 95	0. 112	7. 0	0. 057	0. 143
1. 95	0. 101	7. 5	0. 052	0. 133
1. 95	0. 096	8. 0	0. 049	0. 125
1. 95	0. 089	8. 5	0. 046	0. 118
1. 95	0. 085	9. 0	0. 044	0. 111
1. 95	0. 079	9. 5	0. 041	0. 105

(3) (a) Take the ratio of $\left(\dfrac{V_o}{V_i}\right)$ as ordinate; take the reciprocal of the resistance R of the standard resistor, $\left(\dfrac{1}{R}\right)$, as abscissa. Graph the curve of $\left(\dfrac{V_o}{V_i}\right)$ versus $\dfrac{1}{R}$.

(3) (b) Graph the linear region of the curve of $\left(\dfrac{V_o}{V_i}\right)$ versus $\dfrac{1}{R}$ and use the graphic method to get the slope B of the straight line and its relative uncertainty.

$$B = 0.\,434\ \Omega,\ u(B) = 0.\,009\ \Omega,\ \frac{u(B)}{B} = 0.\,21,\ \left(\frac{u(B)}{B}\right)^2 = 0.\,000\,44$$

Remark

$u(B)$ may be calculated by using several methods. As long as it is calculated, the resulting values closing to the correct value are recognized to

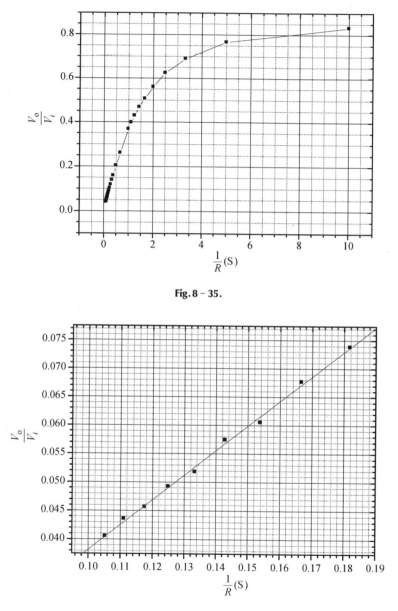

Fig. 8 – 35.

Fig. 8 – 36.

be correct.

(4) With the give length $L = (30.500 \pm 0.025)$ mm and diameter of the

liquid cylinder $d = (13.900 \pm 0.025)$mm, calculate K and its relative uncertainty.

$$K = \frac{1}{B}\frac{L}{S} = \frac{4 \times 30.50}{0.434 \times 13.90^2 \times 3.142} = 0.463(S/mm)$$

$$\left(\frac{u(K)}{K}\right)^2 = \left(\frac{u(B)}{B}\right)^2 + \left(\frac{u(L)}{L}\right)^2 + \left(2 \times \frac{u(d)}{d}\right)^2$$

$$= 0.000\ 44 + 0.000\ 001 + 0.0013$$

$$= 0.0017.$$

(5) Measure the conductivity of the liquid in the container and write the result.

With the given experimental apparatus measure the conductivity of the liquid, the formulae for calculating the liquid conductivity and its relative uncertainty are

$$\sigma = \left(\frac{1}{B}\frac{L}{S}\right)\frac{V_o}{V_i} = K \times A = 0.463 \times A(S/mm),$$

$$A = \frac{V_o}{V_i}, \left(\frac{u(\sigma)}{\sigma}\right)^2 = \left(\frac{u(K)}{K}\right)^2 + \left(\frac{u(A)}{A}\right)^2.$$

Repeat the measurement of $\frac{V_o}{V_i}$ for six times. The resulting data suggested are listed in the table below:

$V_i(V)$	$V_o(V)$	$\frac{V_o}{V_i}$	$V_i(V)$	$V_o(V)$	$\frac{V_o}{V_i}$
1.95	0.037	0.0190	1.95	0.037	0.0190
1.95	0.037	0.0190	1.95	0.038	0.0195
1.95	0.037	0.0190	1.95	0.038	0.0195

$$A = \frac{V_o}{V_i} = 0.019\ 17,$$

$$\sigma = 0.463 \times 0.019\ 17 = 0.008\ 88(S/mm),$$

$$u(A) = \sqrt{\frac{\sum_{i=1}^{6}(A - A_i)^2}{6}} = 0.000\ 258, \frac{u(A)}{A} = 0.013,$$

$$\left(\frac{u(A)}{A}\right)^2 = 0.00017, \quad \left(\frac{u(\sigma)}{\sigma}\right)^2 = 0.001\,7 + 0.000\,17 = 0.001\,9$$

$$\frac{u(\sigma)}{\sigma} = 0.044$$

$$u(\sigma) = 0.000\,39(\text{S/mm}).$$

Therefore the measured conductivity of the liquid is

0. 008 88 ± 0. 000 39(S/mm) or 0. 0089 ± 0. 0004(S/mm).

Remark

The above experimental data are obtained with a homogeneous solution after stirring, and the solute is salt (NaCl, 100 mL), while the solvent is water (700 mL, 10. 1℃).

Appendices

Statutes

1

In recognition of the growing significance of physics in all fields of science and technology, and in the general education of young people, and with the aim of enhancing the development of international contacts between countries of the Asian region in the field of school education in physics, an annual physics competition has been organized for high school students; the competition is called the "Asian Physics Olympiad" and is a competition between individuals. By the term "countries of the Asian region" one should understand countries whose capitals are localized in the region traditionally recognized as Asian. The Asian Physics Olympiad should be conducted not later than two months prior to the International Physics Olympiad.

2

The competition is organized by the Education Ministry or another appropriate institution of one of the participating countries on whose territory the competition is to be conducted. Hereunder, the term "Education Ministry" is used in the above meaning. The organizing country is obliged to ensure equal participation of all the delegations, and to invite all the participants of any of the last three competitions. Additionally, it has the right to invite other countries.

The Asian Physics Olympiad is a purely educational event. No country may have its team excluded from participation on any political grounds resulting from political tensions, lack of diplomatic relations, lack of recognition of some country by the government of the organizing country, imposed embargoes and similar reasons. When difficulties preclude formal invitation of the team representing a country, students from such a country should be invited to participate as individuals.

Within five years of its entry in the competition a country should declare its intention to be the host for a future Olympiad. This declaration should propose a timetable so that a provisional list of the order of countries willing to arrange Olympiads can be compiled.

A country which refuses to organize the competition may be barred from participation, even if delegation from that country has taken part in previous competitions. Any kind of religious or political propaganda against any other country at the Olympiad is forbidden. A country which violates this rule may be barred from participation.

3

The Education Ministries of the participating countries, as a rule, assign the organization, preparation and execution of the competition to a physics society or another institution in the organizing country.

The Education Ministry of the organizing country notifies the Education Ministries of the participating countries of the name and address of the institution assigned to the organization of the competition.

4

Each participating country sends a team consisting of students of general or technical high schools, i. e. schools which cannot be considered technical colleges. Also students who finished their school examination in the year of the competition can be members of a team as long as they do not start the university studies. The age of the participants should not exceed twenty on June 30th of the year of the competition. Each team should normally have 8 members. In addition to the students, two accompanying persons are invited from each country, one of whom is designated delegation head (responsible for whole delegation), and the other — pedagogical leader (responsible for the students). The accompanying persons become members of the

International Board, where they have equal rights. The members of the International Board are treated as contact persons in participating countries on the Asian Physics Olympiad affairs until the next competition.

The competition is conducted in the friendly atmosphere designed to promote future collaborations and to encourage the formation of friendships in the scientific community. To that effect all the possible political tensions between the participants should not be reflected in any activity during the competition any political activity directed against any individuals or countries is strictly prohibited.

The delegation head and pedagogical leader must be selected from specialists in physics or physics teachers, capable of solving the problems of the competition competently. Normally each of them should be able to speak English.

The delegation head of each participating team should, on arrival, hand over to the organizers a list containing personal data on the contestants (given name, family name, date of birth, home address, type and address of the school attended).

5

The working language of the Asian Physics Olympiad is English. Also the competition problems and their solutions should be prepared in English; the organizers, however, may prepare those documents in other languages as well.

6

The financial principles of the organization of the competition are as follows: The Ministry which sends the students to the competition covers the return travel costs of the students and the accompanying persons to the place at which the competition is held. All other costs from the moment of arrival until the moment of departure are covered by the Ministry of the organizing country. In particular, this concerns the costs for board and lodging for the students and the accompanying persons, the costs for excursions, awards for the winners, etc.

7

The competition is conducted on two days, one for the theoretical competition and one for the experimental competition. There should be at least one day of rest between these two days. The time allotted for solving the problem should normally be five hours. The number of theoretical problems should be three and the number of experimental problems one or two.

When solving the problems the contestants may make use of tables of logarithms, tables of physical constants, slide-rules, non-programmable pocket calculators and drawing material. These aids will be brought by the student themselves. Collections of formulae from mathematics or physics are not allowed.

The theoretical problems should involve at least four areas of physics taught at high school level (see Appendix). High School students should be able to solve the competition problems with standard high school mathematics and without extensive numerical calculation.

The host country has to prepare one spare problem which will be presented to the International Board if one of the first three theoretical problems is rejected by two thirds of members of the International Board. The rejected problem cannot be considered again.

8

The competition tasks are chosen and prepared by the host country.

9

The marks available for each problem are defined by the organizer of the competition, but the total number of points for the theoretical problems should be 30 and for the experimental 20. The laboratory problems should consist of theoretical analysis (plan and discussion) and experimental execution.

The winners will receive diplomas or honorable mentions in accordance with the number of points accumulated as follows:

The mean number of points accumulated by the three best participants is considered as 100%.

The contestants who accumulate more than 90% of points receive first prize (diploma).

The contestants who accumulate more then 78% up to 89% receive second prize (diploma).

The contestants who accumulate more than 65% up to 77% receive third prize (diploma).

The contestants who accumulate more than 50% up to 64% receive an honorable mention.

The contestants who accumulate less than 50% of points receive certificates of participation in the competition.

The mentioned marks corresponding to 90%, 78%, 65% and 50% should be calculated by rounding off to the nearest lower integers. The participant who obtains the highest score will receive a special prize and diploma.

Special prizes can be awarded.

10

The obligations of the organizer:

The organizer is obliged to ensure that the competition is conducted in accordance with the Statutes.

The organizer should produce a set of "Organization Rules", based on the Statutes, and send them to the participating countries in good time. These Organization Rules shall give details of the Olympiad not covered in the Statutes, and give names and addresses of the institutions and persons responsible for the Olympiad.

The organizer establishes a precise program for the competition (schedule for the contestants and the accompanying persons, program of excursions, etc.), which is send to the participating countries in advance.

The organizer should check immediately after the arrival of each delegation whether its contestants meet the conditions of the competitions.

The organizer chooses (according to #7 and the list of physics contents in the Appendix to these Statutes) the problems and ensures their proper formulation in English and in other languages set out in #5. It is advisable to select problems where the solutions require a certain creative capability and a considerable level of knowledge. Everyone taking part in the preparation of the competition problems is obliged to preserve complete secrecy.

The organizer must provide the teams with interpreters.

The organizer should provide the delegation leaders with photostat copies of the solutions of the contestants in their delegation before the final classification.

The organizer is responsible for the grading of the problem solutions.

The organizer drafts a list of participants proposed as winners of the prizes and honorable mentions.

The organizer prepares the prizes (diplomas), honorable mentions and awards for the winners of the competition.

The organizer is obliged to publish proceedings (in English) of the Olympiad. Each of the participants of the competition (delegation heads, pedagogical leaders and contestants) should receive one copy of the proceedings free of charge not later than one year after the competition.

11

The scientific part of the competition must be within the competence of the International Board, which includes the delegation heads and pedagogical leaders of all the delegations.

The Board is chaired by a representative of the organizing country. He is responsible for the preparation of the competition and serves on the Board in addition to the accompanying persons of the respective teams.

Decisions are passed by a majority vote. In the case of equal number of votes for and against, the chairman has the casting vote.

12

The delegation leaders are responsible for the proper translation of the problems from English (or other languages mentioned in #5) to the mother tongue of the participants.

13

The International Board has the following responsibilities:

to direct the competition and supervise that it is conducted according to the regulations;

to ascertain, after the arrival of the competing teams, that all their members meet the requirements of the competition in all aspects. The Board will disqualify those contestants who do not meet the stipulated conditions. The costs incurred by a disqualified contestant are covered by his country;

to discuss the Organizers' choice of tasks, their solutions and the suggested evaluation guidelines before each part of the competition. The Board is authorized to change or reject suggested tasks but not to propose new ones. Changes may not affect experimental equipment. There will be a final decision on the formulation of tasks and on the evaluation guidelines. The participants in the meeting of the International Board are bound to preserve secrecy concerning the tasks and to be of no assistance to any of the participants;

to ensure correct and just classification of the prize winners; the grading of those contestants who do not receive prizes or honorable mentions is not to be disclosed;

to establish the winners of the competition and make a decision concerning presentation of the prizes and honorable mentions. The decision of the International Board is final to review the results of the competition;

to select the country which will be assigned the organization of the next competition.

The International Board is the only body having the right to take decisions on barring countries from participation in the Asia Physics

Olympiads for violation of these Statutes.

Observers may be present at the meetings of the International Board, but not to vote or take part in the discussion.

14

The institution in charge of the Olympiad announces the results and presents the awards and diplomas to the winners at an official gala ceremony. It invites representatives of the organizing Ministry and scientific institutions to the closing ceremony of the competition.

15

The long-term work involved in organizing the Olympiads is coordinated by a "Secretariat for the Asia Physics Olympiads". This Secretariat consists of the President and Secretary. They are elected by the International Board for a period of five years when the chairs become vacant. For special merits to the Asia Physics Olympiads the International Board may award one person with the lifelong title "Honorable President of the Asia Physics Olympiads" and a number of persons with the lifelong title "Honorable Member of the International Board of the Asia Physics Olympiad". The Honorable President and the Honorable Members of the International Board of the Asia Physics Olympiads are members of the International Board in addition to the regular members mentioned in ♯ 4. They are invited to each Asia Physics Olympiad at cost (including travel expenses) of the organizing country.

16

The present Statutes have been drafted on the basis of the Statutes of the International Physics Olympiads.

Changes in these Statutes, the insertion of new paragraphs or exclusion of old ones, can only be made by the International Board and requires qualified majority ($\frac{2}{3}$ of the votes).

No changes may be made to these Statutes or Syllabus unless each delegation obtained written text of the proposal at least three months in advance.

17

Participation in an Asia Physics Olympiad signifies acceptance of the present Statutes by the Education Ministry of the participating country.

18

The originals of these Statues are written in English.

Syllabus

General

a) The extensive use of the calculus (differentiation and integration) and the use of complex numbers or solving different equations should not be required to solve the theoretical and practical problems.

b) Questions may contain concepts and phenomena not contained in the Syllabus but sufficient information must be given in the questions so that candidates without previous knowledge of these topics would not be at a disadvantage.

c) Sophisticated practical equipment likely to be unfamiliar to the candidates should not dominate a problem. If such devices are used then careful instructions must be given to the candidates.

d) The original texts of the problems have to be set in the SI units.

A. Theoretical Part

The first column contains the main entries while the second column contains and remarks if necessary.

1. Mechanics	
a) Foundation of kinematics of a point mass.	*Vector description of the position of the point mass velocity and acceleration as vectors.*
b) Newton's laws, inertial systems.	*Problems may be set on changing mass.*
c) Closed and open systems, momentum and energy, work, power.	
d) Conservation of energy, conservation of linear momentum, impulse.	
e) Elastic forces, frictional forces, the law of gravitation, potential energy and work in a gravitational field.	*Hooke's law, coefficient of fiction $\left(\frac{F}{R} = const \right)$, fictional forces static and kinetic, choice of zero of potential energy.*
f) Centripetal acceleration, Kepler's law.	

Cont.

2. Mechanics of Rigid Bodies	
a) Statics, center of mass, torque.	*Couples, conditions of equilibrium of bodies.*
b) Motion of rigid bodies, translation, rotation, angular velocity, angular acceleration, conservations of angular momentum.	*Conservation of angular momentum about fixed axis only.*
c) External and internal forces, equation of motion of a rigid body around the fixed axis, moment of inertia, kinetic energy of a rotating body.	*Parallel axes theorem (Steiner's theorem), additivity of the moment of inertia.*
d) Accelerated reference systems, inetial forces.	*Knowledge of the Coriolis force formula is not required.*

3. Hydromechanics
No specific questions will be set on this but students would be expected to know the elementary concepts of pressure, buoyancy and the continuity law.

4. Thermodynamics and Molecular Physics	
a) Internal energy, work and heat, first and second laws of thermodynamics.	*Thermal equilibrium, quantities depending on state and quantities depending on process.*
b) Model of a perfect gas, pressure and molecular kinetic energy, Avogadro's number, equation of state of a perfect gas, absolute temperature.	*Also molecular approach to such simple phenomena in liquids and solids as boiling, melting etc.*
c) Work done by an expanding gas limited to isothermal and adiabatic processes.	*Proof of the equation of the adiabatic process is not required.*
d) The Carnot cycle, thermodynamic efficiency, reversible and irreversible processes, entropy (statistical approach), Boltzmann factor.	*Entropy as a path independent fiction, entropy changes and reversability, quasistatic processes.*

5. Oscillations and Waves	
a) Harmonic oscillations, equation of harmonic oscillation.	*Solution of the equation for harmonic motion, attenuation and resonance-qualitatively.*

Cont.

b) Harmonic waves, propagation of waves, transverse and longitudinal waves, linear polarization, the classical Doppler effect, sound waves.	*Displacement in a progressive wave and understanding of graphical representation of the wave, measurements of velocity of sound and light, Doppler effect in one dimension only, propagation of waves in homogeneous and isotropic media, reflection and refraction, Fermats principle.*
c) Superposition of harmonic waves, coherent waves, interference, beats, standing waves.	*Realization that intensity of wave is proportional to the square of its amplitude. Fourrier analysis is not required but candidates should have some understanding that complex waves* *can be made from addition of simple sinusoidal waves of different frequencies. Interference due to thin films and other simple systems (final formulae are not required),* *superposition of waves from secondary sources (diffraction).*
6. Electric Charge and Electric Field	
a) Conservation of charge, Coulomb's law.	
b) Electric field, potential, Gauss' law.	*Gauss' low confined to simple symmetric systems like sphere, cylinder, plate, etc, electric dipole moment.*
c) Capacitors, capacitance, dielectric constant, energy density of electric field.	
7. Current and Magnetic Field	
a) Current, resistance, internal resistance of source, Ohm's law, Kirchhoff's laws, work and power of direct and alternating currents, Joule's law.	Simple cases of circuits containing non-ohmic devices with known V-I characteristics.
b) Magnetic field (B) of a current, current in a magnetic field, Lorentz force.	Particles in a magnetic field, simple applications like cyclotron, magnetic dipole moment.
c) Ampere's law.	Magnetic field of simple symmetric systems like straight wire, circular loop and long solenoid.
d) Law of electromagnetic induction, magnetic flux, Lenz's law, self-induction, inductance, permeability, energy density of magnetic field.	

Cont.

e) Alternating current, resistors, inductors and capacitors in AC-circuits, voltage and current (parallel and series) resonance.	Simple AC-circuits, time constants, final formulae for parameters of concrete resonance circuits are not required.

8. Electromagnetic Waves

a) Oscillatory circuit, frequency of oscillations, generation by feedback and resonance.

b) Wave optics, diffraction from one and two slits, diffraction grating, resolving power of a grating, Bragg reflection.

c) Dispersion and diffraction spectra, line spectra of gases.

d) Electromagnetic waves as transverse waves, polarization by reflection, polarizers.	Superposition of polarized waves.

e) Resolving power of imaging systems.

f) black body, Stefan-Boltzmann's law.	Planck's formula is not required.

9. Quantum Physics

a) Photoelectric effect, energy and impulse of the photon.	Einstein's formula is required.

b) De Broglie wavelength, Heisenberg's uncertainty principle.

10. Relativity

a) Principle of relativity, addition of velocities, relativistic Doppler effect.

b) Relativistic equation of motion, momentum, energy, relation between energy and mass, conservation of energy and momentum.

11. Matter

a) Simple applications of the Bragg equation.

b) Energy levels of atoms and molecules (qualitatively), emission, absorption, spectrum of hydrogen-like atoms.

c) Energy levels of nuclei (qualitatively), alpha-, beta-and gamma-decays, absorption of radiation, half life and exponential decay, components of nuclei, mass defect, nuclear reactions.

B. Practical Part

The Theoretical Part of the Syllabus provides the basis for all the experimental problems. The experimental problems given in the experimental contest should contain measurements.

Additional requirements:

Candidates must be aware that instruments affect measurements.

Knowledge of the most common experimental techniques for measuring physical quantities mentioned in Part A.

Knowledge of commonly used simple laboratory instruments and devices such as calipers, thermometers, simple volt-, ohm- and ammeters, potentiometers, diodes, transistors, simple optical devices and so on.

Ability to use, with the help of proper instruction, some sophisticated instruments and devices such as double-beam oscilloscope, counter, ratemeter, signal and function generators, analog-to-digital converter connected to a computer, amplifier, integrator, differentiator, power supply, universal (analog and digital) volt-, ohm- and ammeters.

Proper identification of error sources and estimation of their influence on the final result(s).

Absolute and relative errors, accuracy of measuring instruments, error of a single measurement, error of a series of measurements, error of a quantity given as a function of measured quantities.

Transformation of a dependence to the linear form by appropriate choice of variables and fitting a straight line to experimental points.

Proper use of the graph paper with different scales (for example polar and logarithmic papers).

Correct rounding off and expressing the final result(s) and error(s) with correct number of significant digits.

Standard knowledge of safety in laboratory work. (Nevertheless, if the experimental set-up contains any safety hazards the appropriate warnings should be included in the text of the problem.)

Team List

Team	APhO							
	1st	2nd	3rd	4st	5st	6st	7st	8th
Australia	G	G	P	P	P		P	P
Azerbaijan						P	P	P
Brunei	O							P
Cambodia					P	P	P	P
China	P		P		P	P	P	H
Chinese Taipei	P	H	P	P	P	P	P	P
Hong Kong, China								P
Georgia			P				P	
India	O	P					P	P
Indonesia	H	P	P	P	P	H	P	P
Israel		P	P	P	P		P	P
Japan		O						O
Jordan		P				P	P	
Kazakhstan	P	P	P		P	P	H	P
Kyrgyzstan			P	P		P	P	P
Laos				P		P		P
Macau, China								P
Malaysia	O	P	P		P	P		
Mongolia		P	P		P		P	P
Nepal								P
Pakistan			P					
Philippines	P			P		P		
Qatar		O				P		
Russia						P		

Cont.

Team	APhO							
	1st	2nd	3rd	4st	5st	6st	7st	8th
Singapore	P	P	H		P	P	P	P
South Korea			O					
Sri Lanka								P
Tajikistan						P	P	P
Thailand	P	P	P	H	P	P	P	P
Turkey			P					
Turkmenistan					P		P	P
Uzbekistan	P		P					
Vietnam	P	P	P	P	H	P	P	P

Note: P for Participation, H for Host, O for Observer, G for Guest